30- SECOND
THEORIES

30-SECOND
THEORIES

The 50 most thought-provoking
theories in science, each explained
in half a minute

Editor
Paul Parsons

Foreword
Martin Rees

Contributors
Jim Al-Khalili
Susan Blackmore
Michael Brooks
John Gribbin
Christian Jarrett
Robert Matthews
Bill McGuire
Mark Ridley

Published in the UK in 2010 by
Icon Books Ltd
Omnibus Business Centre
39–41 North Road, London N7 9DP
email: info@iconbooks.co.uk
www.iconbooks.co.uk

This book was conceived, designed,
and produced by
Ivy Press
The Old Candlemakers
West Street, Lewes,
East Sussex BN7 2NZ, UK
www.ivy-group.co.uk

Creative Director **Peter Bridgewater**
Publisher **Jason Hook**
Editorial Director **Caroline Earle**
Art Director **Michael Whitehead**
Commissioning Editor **Nic Compton**
Designers **James Hollywell, Les Hunt**
Concept Design **Linda Becker**
Illustrations **Jon Raimes**
Glossaries & Profiles Text **Tom Jackson**
Picture Research **Lynda Marshall**

ISBN: 978-1-848311-29-9

Printed and bound in China

10 9 8 7 6 5 4 3 2 1

CONTENTS

FOREWORD
Martin Rees

Our world is getting ever more complex, more baffling. Some pessimists argue that scientific progress – or indeed society itself – will clog up because of 'information overload'. I don't think that's a serious worry. As science advances, more patterns and regularities are revealed in nature. These advances cut down the number of disconnected facts worth remembering. There's no need to record the fall of every apple, because, thanks to Isaac Newton, we understand how gravity pulls everything – whether apples or spacecraft – towards Earth.

The simplest building blocks of our world – atoms – behave in ways we can understand and calculate. And the laws and forces governing them are universal: atoms behave the same way everywhere on Earth – indeed, they are the same, even in the remotest stars. We know these basic facts well enough to enable engineers to design all the mechanical artefacts of our modern world, from radios to rockets.

Our everyday environment is too complicated for its essence to be captured by a few formulae. But our perspective on our Earth has been transformed by great, unifying ideas. The concept of continental drift, for instance, helps us to fit together a whole raft of geological and ecological patterns across the globe. Charles Darwin's great insight – evolution by natural selection – reveals the overarching unity of the entire web of life on our planet. Whatever our personal lives may be like, our environment is neither chaotic nor anarchic. There are patterns in nature. There are even patterns in how we humans behave – in how cities grow, how epidemics spread, and how technologies, such as computer chips, develop. The more we understand the world, the less bewildering it becomes, and the more we're able to change it.

These laws or patterns are the great triumphs of science. To discover them has required dedicated talent – even genius in many cases. But to grasp their essence is not so difficult. We all appreciate music, even if we can't compose or perform it. Likewise, the ideas of science can be accessed and marvelled at by everyone.

Science impinges more than ever on our lives. Many political issues – energy, health, environment and so forth – have a scientific dimension. How science is applied matters to us all. The important choices shouldn't be made just by scientists; they should be the outcome of wider public debate. But in order for that to happen, we all need a 'feel' for the key ideas of science. And, quite apart from their practical uses, these ideas should be part of our common culture. The great concepts of science can be conveyed briefly – maybe even in 30 seconds – using non-technical words and simple images. That's the aim of this book, and we should hope it succeeds.

The theory of everything

Unification theory, one example of which is string theory, attempts to explain how everything in the universe is connected. Great scientists have spent years trying to develop a 'theory of everything' – this book explains that research in just 30 seconds (see page 50).

Testing, testing

Unlike the pet theories that we all like to make up, scientific theories are supported by cold hard evidence, usually in the form of carefully planned and controlled experiments.

INTRODUCTION
Paul Parsons

Everyone's got their own pet theory. I should
know. During my time as editor of the monthly BBC science and
technology magazine *Focus*, the postbag brought several of them every
day – missives from readers claiming to have cracked the mysteries of
black holes, parallel universes or the Big Bang; determined the origin
of life; or unified the laws of particle physics. I'd reply, thanking them
for their theories, and requesting that they send in the full supporting
mathematics. I don't think any of them ever did.

That's the difference between the 'theories' we bandy about in
everyday parlance – our inklings and just-thought-of guesstimates –
and the theories that are painstakingly constructed by scientists.

A theory in science is a logical creation. It reflects the most accurate
experimental observations and the best understanding of how the world
works. Yet a scientific theory doesn't necessarily represent absolute
truth. It can only capture the state of our knowledge so far. There's every
chance that a new piece of evidence will come to light that disproves the
theory, and sends the theoreticians back to the drawing board.

One example of this is our view of the Solar System. In the second
century AD, the Greek philosopher Ptolemy developed the theory that the
Earth lies at the centre of the Solar System – a sound explanation for
the primitive astronomical observations of the day. But, in the early 17th
century, the Italian astronomer Galileo began to survey the skies with the
newly invented telescope. It permitted observations of the Solar System
that were vastly superior to anything achieved with the naked eye.

Galileo's observations revealed details that fitted with a new theory,
developed by the Polish astronomer Nicolaus Copernicus a hundred years
earlier. Copernicus' theory painted the now-familiar picture of the Sun,
not the Earth, sitting at the heart of our Solar System. Many observations
since – including data from space probes – have confirmed the Sun-
centred view of the Solar System.

Other casualties include the flat Earth theory, phlogiston theory – an early attempt to explain the origin of fire – and so-called intelligent design theory. Our theoretical understanding of pretty much every branch of modern science has evolved in this way, with old, defunct theories being replaced by new and improved ones.

The theories composing the body of scientific knowledge today cover everything from the origin of the Universe to the workings of the human mind. Over the pages that follow, the 50 greatest theories are laid out by some of the most talented science communicators. Each is summed up in a single, user-friendly passage encapsulating its essence. No jargon, no waffle – just concise, plain English.

The theories are organized into seven pillars of understanding. The first is The Macrocosm, and deals with the large-scale physics of the everyday world, such as the laws of motion, gravity and electricity. The Microcosm turns our attention to the very small, looking at the quantum world of atoms and other subatomic particles of nature. The third pillar focuses on Human Evolution – how life, people and facets such as intelligence and language all came to be. Mind & Body charts key theories in medicine – from psychoanalysis to gene therapy. In Planet Earth, we survey the great theories that have enabled scientists to grasp the inner workings of our planet and its climate. The Universe casts an eye further afield, taking stock of the origin, evolution and ultimate fate of our cosmos – and others. The final pillar, The Knowledge, deals with branches of science concerned with the growth of science itself, such as Moore's Law for the constant improvement in the power of computers and Ockham's Razor – quite literally, the mother of all theories. The pillars also include profiles of some of the giants in these fields – summarizing the lives of intellectual heavyweights from Charles Darwin to Stephen Hawking.

This book serves a dual purpose. Its structured, piecemeal approach makes it an excellent reference to dip into as required – a mini-encyclopedia of theoretical science. On the other hand, read it cover to cover, and you'll have an excellent overview of how scientists today think the natural world works. So, if you're in a quandary about quantum theory or wrestling with relativity, or just curious about what exactly scientists have been up to all these years, then sit back in your favourite armchair and let our resident experts guide you through the greatest achievements of the human mind. But please – keep your pet theories to yourselves!

Relative values

*The theory of relativity is probably one of the
best-known scientific theories – but do we really
understand it? It's all about how time, matter, energy
and space interact (apparently) – see page 30 for the
half-minute explanation.*

THE MACROCOSM

THE MACROCOSM
GLOSSARY

atom The smallest unit of any substance found on Earth. Atoms themselves are made up of yet smaller particles: protons, neutrons and electrons. The precise combination of these particles gives each type of atom its physical and chemical properties. For example, a gold atom has a different make up from an atom of carbon.

constant A physical quantity that is measured in nature and does not change. One such constant is the speed of light. Constants can be used to relate one physical property to another, properties that are said to be 'proportional' to one another. When one of them changes, the other also changes by an equal proportion. The constant allows you to calculate exactly how much one change will affect the other.

dimension A fundamental measure used to describe an object or event. Humans are aware of four dimensions – length, width, height and time – but scientific theories often involve multiple dimensions that are only perceived through mathematics.

electric charge A basic property of matter. Some matter, such as a proton, is positively charged; other types, such as electrons, are negatively charged. Neutrons, on the other hand, are neutral – they have no charge. An electric current is a flow of electrons (or other charged objects) from a negatively charged object to a positively charged one.

electromagnetic wave Another way of describing radiation, such as light and heat.

equation A mathematical notation used to show how measurable quantities relate to each other. $E=mc^2$ is an equation showing how the energy in an object (E) equals the object's mass (m) multiplied by the speed of light (c) squared (^{2}). (Squared means a number multiplied by itself once.)

field An area of space in which a force has an effect on matter. Examples include magnetic and gravitational fields.

kinetic energy The energy contained in a moving body that relates to its motion.

law A simple description of a pattern that has been observed in nature. Most laws are expressed as equations.

macrocosm The big picture – a model that reflects the functioning of a system on the largest of scales.

mass A measure of the quantity of matter in an object. 'Mass' and 'weight' are often used interchangeably, but weight is really a measure of the pull of gravity on the object. In everyday terms, the 'mass' and 'weight' of an object are effectively the same on Earth, but on the Moon, the same object's mass is unchanged, while its weight is reduced by 85 per cent by the lower gravity.

matter The stuff of the Universe, which fills space and can be measured in some way.

oscillations Rhythmic movements which occur around a central, unchanging point in space.

particles Small units of matter. In physics on the tiniest length scales, a particle may be a minute building block within an atom, or a molecule of water, oxygen or any other substance. Otherwise it may be a speck of dust, smoke or sand among many.

potential energy The energy stored within an object that could be released and harnessed to do useful work. A boulder teetering on a hilltop has potential energy. If it is pushed down the hill, that potential will be converted to kinetic, or movement, energy.

perpendicular At a right angle – 90 degrees – to something else. Walls are perpendicular to the floor – hopefully.

radiation A term sometimes used to describe the dangerous emissions from radioactive substances, but more correctly used to describe the transfer of photons – tiny packets of energy – through space. Light, heat, radio waves, as well as dangerous gamma rays, are all types of radiation, each carrying varying amounts of energy.

refraction This is when a beam of light, or other radiation, changes direction slightly as it is passed from one medium (e.g. the air) to another (e.g. water). The refraction is due to the differences in the speed of light in the two media. If a beam of light arrives at the interface between the two media at an angle, one side of the beam will change speed before the other side. As a result the whole beam changes direction slightly.

speed of light The speed at which radiation travels, and the speed limit of the Universe. The speed of light in a vacuum is 299,792 kilometres per second (186,282 miles per second). Nothing can travel faster than this.

subatomic Smaller than an atom.

PRINCIPLE OF LEAST ACTION

the 30-second theory

RELATED THEORIES
see also
UNIFICATION
page 50
OCKHAM'S RAZOR
page 142

3-SECOND BIOGRAPHIES
LEONHARD EULER
1707–1783
PIERRE DE FERMAT
1601–1665
GOTTFRIED LEIBNITZ
1646–1716
VOLTAIRE
1694–1778

30-SECOND TEXT
Michael Brooks

3-SECOND THRASH
At the core of modern physics is the notion that 'Nature is thrifty in all its actions ...'

3-MINUTE THOUGHT
Quantum theory, which describes how things work on the subatomic scale, seems to be the one area where the principle of least action does not apply. Quantum objects can be in two states at once, and can take multiple paths when travelling from one place to another. Richard Feynman went so far as to suggest that a quantum particle will simultaneously take every possible path when making a journey!

This says, essentially, that things happen in the way that requires least effort. So, a beam of light will travel in a straight line because that is the shortest path between two points. If you drop a ball, it will travel towards the centre of the Earth. No one is quite sure who came up with the principle of least action, but your everyday experience would probably lead you to come up with it if you thought about it for a bit. In the 18th century, though, this was a big deal. Some of the greatest names in mathematics, such as Leonhard Euler, Pierre de Fermat, Gottfried Leibnitz, and Voltaire were involved in the argument over who came up with the idea first. It was important to make these kinds of statements at the time, because they led to the formation of the equations that describe how things move when acted on by forces. They also led to the concepts of potential and kinetic energy.

As theories go, the principle of least action is just common sense: natural motion always takes the easiest and shortest route.

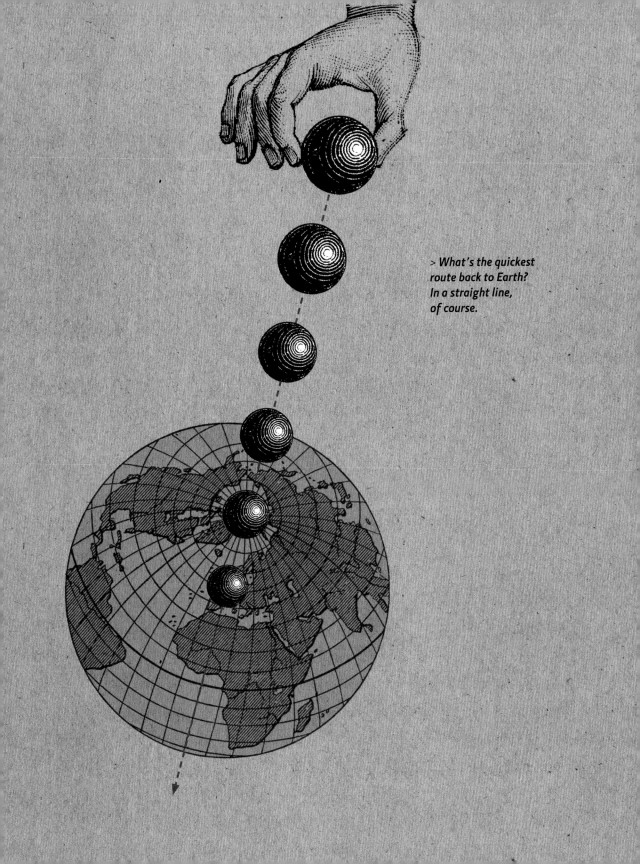

> *What's the quickest route back to Earth? In a straight line, of course.*

LAWS OF MOTION

the 30-second theory

When Isaac Newton sat down

and thought about how things move, he worked out three laws that are now so familiar they seem like common sense. First, he said that objects have 'inertia', which is a measure of resistance to changes in their motion. Inertia means that things remain still until you give them a push. Similarly, objects that are moving keep moving unless something stops or pushes on them. Second, the mass of the object determines what effect a particular push will have on the motion (or lack of it). The third law, which is the most famous, feels slightly different. It says that every action has an equal and opposite reaction. If I push you, I feel an equal push in return. This is the principle by which space rockets and jet engines work: when they push out an exhaust gas from the nozzle at the rear, the engines get a push forward. This is why you should be careful when you step off a boat. To move yourself forward, you inevitably move the boat backward. If you don't take that into account, you can end up taking a swim!

3-SECOND THRASH
Newton formulated the basic description of how things move and effectively invented rocket science.

3-MINUTE THOUGHT
Newton's laws are simple but powerfully accurate. They are not accurate enough, however, to describe what happens when things are moving at close to the speed of light, or in strong gravitational fields. In such instances, Einstein's theory of relativity takes over and provides our ultimate laws of motion.

RELATED THEORIES
see also
UNIVERSAL GRAVITATION THEORY
page 20
THEORY OF RELATIVITY
page 30
UNIFICATION
page 50

3-SECOND BIOGRAPHY
ISAAC NEWTON
1643–1727

30-SECOND TEXT
Michael Brooks

The laws of motion are all you need to describe how everyday objects move – from footballs to space stations. Newton gave us the means to plan a journey to the Moon – it just took 300 years to invent the rockets needed to do it.

> *Fly me to the Moon – using Newton's laws of motion.*

UNIVERSAL GRAVITATION THEORY

the 30-second theory

3-SECOND THRASH
What goes up must come down – and it will do so, just as Newton said it should.

3-MINUTE THOUGHT
Some ideas in modern physics suggest that Newton's law of gravitation may need adjusting to consider things separated by less than a millimetre, or more than the diameter of the Solar System. What's more, no one has a good explanation for why things with mass attract each other in the first place, why gravity is much weaker than the other forces of nature, or for the true value of the gravitational constant, which is the least accurately measured constant of physics.

This description of one of the fundamental forces of nature is among the greatest achievements in science. Isaac Newton came up with it in 1687 as part of his masterful *Principia Mathematica*, a three-volume description of mathematics. Universal gravitation theory says that there is a mutual attraction between anything that has mass – anything made of normal matter, that is. That attraction depends on the two masses involved, the distance between them, and a constant known as the gravitational constant. One of the central insights of the theory was that the gravitational force follows an 'inverse square law'. This means the attraction between the two objects diminishes as the square of the distance between them. Newton's formulation of the law was so accurate that it immediately explained the motion of the planets, creating an easy way to predict their movements relative to each other and the Sun. It has also enabled us to send rockets into space. After Einstein came up with the theory of relativity and used it to explain some small anomalies in the planetary orbits, it was realized that Newton's law was not quite the final word on gravity. However, it is almost universally accurate when applied to the gravitational attractions we encounter in everyday life.

RELATED THEORIES
see also
LAWS OF MOTION
page 18
THEORY OF RELATIVITY
page 30
QUANTUM FIELD THEORY
page 46
UNIFICATION
page 50

3-SECOND BIOGRAPHY
ISAAC NEWTON
1643–1727

30-SECOND TEXT
Michael Brooks

Large or small, everything comes down to Earth with the same bump.

$$F=G \ \frac{m_1 \times m_2}{r^2}$$

> The acceleration force due to gravity is the same for a massive elephant as it is for a tiny pea – but get out of the way of falling elephants!

=

1879
Born, Ulm, Germany

1896
Attends college in Zurich
to train as a physics and
mathematics teacher

1905
Publishes four papers on
light, molecular motion
and energy

1913
Begins work on his new
theory of gravity

1915
Completes theory of
general relativity

1921
Wins Nobel Prize
for Physics

1928
Begins work on his
unified field theory

1935
Moves to the
United States

1955
Dies, Princeton,
United States

ALBERT EINSTEIN

'What would you see if you sat on a beam of light?' That was the question Albert Einstein asked himself when still a boy. The answer Einstein came up with – the theory of relativity – broke apart the ordered Universe described by Sir Isaac Newton 250 years before. Nearly a century after Einstein, today's physicists are still trying to figure out just what exactly his theory revealed.

Albert Einstein was born in southern Germany in 1879. His father Hermann was an unsuccessful businessman. When the family fortunes took a turn for the worse in the 1890s, Albert was left to finish his education alone in Germany, while his parents worked in Italy. By this time, Albert had already begun his own scientific research and dropped out of school at the age of 16. Despite his lack of formal schooling, Albert still won a place at Zurich technical college in 1896.

Nevertheless Einstein's attendance record remained poor, as he pursued his own studies at home. After graduation, Einstein's bad reputation barred him from continuing an academic career. He took a job at the patent office in Bern, and married Milevia Maric in 1903. His simple job left him with free time to think about physics.

Einstein's 'miracle year' was 1905. He published four papers. One of them, on the relationship between light and electricity, won him the Nobel Prize. Another was the starting point for the theory of relativity. Ten years later Einstein published his general theory of relativity. This theory brought together his ideas on energy, mass and gravity into a single concept called space-time.

As his life was rocked by world war and personal problems, Einstein continued to work. His main goal was to connect relativity with the theories that governed atoms, thus creating a single unified theory of everything. This work was incomplete upon Einstein's death in 1955 and remains unfinished today.

WAVE THEORY

the 30-second theory

You only have to go to the beach and be hit by a wave to appreciate that waves carry energy. But they do so in surprisingly diverse ways. Some waves, such as sea waves and sound waves, physically move the particles of water, or air, or whatever medium they are travelling through. These waves come in two types. A sound wave is 'longitudinal' – it creates vibrations that move air parallel to the direction in which the wave is moving. 'Transverse' waves, such as electromagnetic waves, oscillate in a direction perpendicular to their direction of travel. Polaroid sunglasses work because they block out transverse oscillations moving in a certain orientation – for example, up and down. Any light waves oscillating in another direction – from side to side, for example – pass through unaffected. If light was a longitudinal wave, polaroid lenses would have no effect at all.

Most of wave theory was worked out in the 19th century when pioneers such as Thomas Young showed how waves can be manipulated. Waves are reflected by certain materials, refracted as they cross the boundary between two media, or diffracted, which means they spread out as they pass through a narrow opening. They can also interfere with each other, cancelling each other out completely in some regions of the medium while combining into larger, more powerful, waves in other areas.

3-SECOND THRASH
The reason why in space no one can hear you scream.

3-MINUTE THOUGHT
With the discovery of quantum theory came the realization that electromagnetic waves are really the movement of packets of energy called photons. Albert Einstein won a Nobel Prize for this discovery. Before then light was thought to be a series of waves, oscillating at different frequencies. The energy held by a photon relates to the frequency of the light wave's oscillation. The energy also relates to the colour we see. Blue light is composed of higher-energy photons – or faster-oscillating waves – than red light.

RELATED THEORIES
see also
ELECTROMAGNETISM
page 28
QUANTUM FIELD THEORY
page 46

3-SECOND BIOGRAPHY
THOMAS YOUNG
1773–1829

30-SECOND TEXT
Michael Brooks

Waves are everywhere: in the ocean, in the air and even in the vacuum of space. Wherever they are, all waves have a wavelength – the distance from the start of one wave cycle to the start of the next.

> If a wave travels one wavelength in a second it has a frequency of 1 Hertz (Hz). An ocean wave might have a frequency of 0.2 Hz, but a light wave has a frequency of about 500 trillion Hz!

transverse wave

wavelength

longitudinal wave

THERMODYNAMICS

the 30-second theory

3-SECOND THRASH
Lord Kelvin's brilliant
description of the nature
of heat shows us that
nothing ever costs
nothing.

3-MINUTE THOUGHT
Until the second law of
thermodynamics came
along, many people
believed it possible
to create 'perpetual-
motion' machines. One
example was designed
to light a house without
using up any energy: An
electric motor turned a
wheel on a generator,
which provided power
for lighting the house
– and also for turning the
electric motor. Although
this may seem daft now,
it was big business in
the 19th century. Many
industrialists tried to solve
the problem, hoping to
make an immense fortune.

If you want to know how heat
moves around, you need to understand
thermodynamics. The theory is governed by
three laws. The first one states that whatever is
going on, the total energy in the Universe stays
the same. In other words, you cannot create
or destroy energy; you can only change one
form of energy into another. The second law
says that an isolated system's entropy always
increases. Entropy is a measure of the part of its
energy that cannot be put to work in some way.
For example, as a watchspring unwinds, it has
less and less power to keep the watch running.
Its entropy rises because the spring's potential
energy is slowly transferred to the hands as
kinetic energy, with some energy also lost as
heat because of friction in the mechanism.

The third law says that, as a system's
temperature drops towards absolute zero
(the lowest possible temperature: $-273.15\,^\circ$C;
$-459.67\,^\circ$F), all natural processes cease to occur,
and the entropy reaches a minimum. The upshot
of this is that it is impossible to reach absolute
zero because no processes can get you there.

Thermodynamics is not as abstract and
esoteric as it sounds. It was developed in the
19th century by Lord Kelvin and forms the
basis for your house's refrigerator and central
heating, the engines that move your car, and
the biological processes that keep you alive.

RELATED THEORIES
see also
CHAOS THEORY
page 152

3-SECOND BIOGRAPHY
WILLIAM THOMSON,
LORD KELVIN
1824–1907

30-SECOND TEXT
Michael Brooks

*According to the theory
of thermodynamics
a refrigerator does
not add cold to your
food, it takes its heat
away. It does this by
compressing a fluid in
the pipes on the back
of the appliance.*

condenser

> An aerosol gets cold when you use it. A refrigerator works in the same way. An expanding liquid becomes a cooling gas – and the compressor then turns the gas back into a warm liquid.

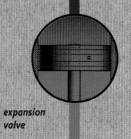

expansion
valve

compressor

expander

ELECTROMAGNETISM

the 30-second theory

Electromagnetism is a stunningly powerful idea. In fact, our lives would be unrecognizable without it. It is all about what happens when you combine electric charges, movement and magnetic fields. Move a metal wire within a magnetic field, and you will cause an electric current to flow in the wire. This is how we generate electricity. Conversely, send an electric current through a wire, and the movement of electric charges will create a magnetic field. This is how we make the electromagnets that power most doorbells and particle accelerators. The third option is to run an electric current through a wire sitting in a magnetic field. The wire will move. This is the idea behind the electric motor in your kitchen whisk and power drill.

The main credit for the theory rests with a Scotsman, James Clerk Maxwell, who first wrote down the equations that describe the complex interplay of electric and magnetic fields. The equations, it turned out, required an unexpected factor: the speed of light. This revelation led to the understanding that light and radiant heat can be thought of as moving variations in electromagnetic fields. These moving fields came to be known collectively as radiation. Investigations of radiation led Max Planck to invent quantum theory and Albert Einstein to come up with his concept of relativity.

RELATED THEORIES
see also
WAVE THEORY
page 24

QUANTUM FIELD THEORY
page 46

UNIFICATION
page 50

3-SECOND BIOGRAPHIES
JAMES CLERK MAXWELL
1831–1879

MAX PLANCK
1858–1947

ALBERT EINSTEIN
1879–1955

30-SECOND TEXT
Michael Brooks

3-SECOND THRASH
A battery, a loop of wire and a magnet can provide a very moving experience!

3-MINUTE THOUGHT
The formulation of quantum theory made it necessary to rewrite James Clerk Maxwell's equations on electricity and magnetism. The result was a new theory called quantum electrodynamics (QED). Intriguingly, though, some parts of the theory had to be fudged with numbers that came from experiments rather than pure theory. Despite this, QED is widely considered to be the most successful theory in science.

It is amazing to think that a flow of invisible particles and the effects of an invisible force field can power everything, from a toy car to a super computer.

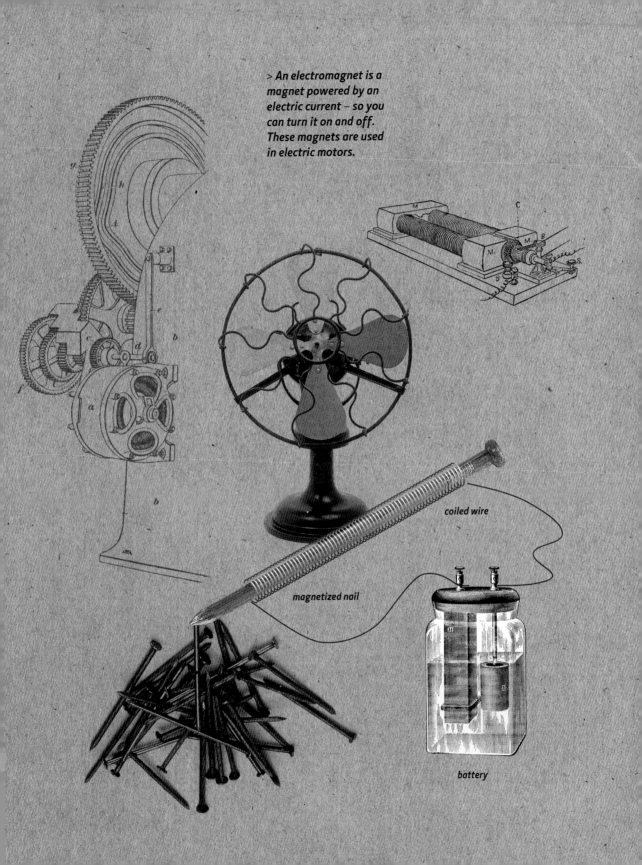

> *An electromagnet is a magnet powered by an electric current – so you can turn it on and off. These magnets are used in electric motors.*

coiled wire

magnetized nail

battery

THEORY OF RELATIVITY

the 30-second theory

3-SECOND THRASH
If you want to stay young, you'd better keep moving!

3-MINUTE THOUGHT
Relativity is at the root of time travel. Astronauts hurtling through space at enormous speeds are the nearest we have got to time travellers. While travelling at high speeds, their time slows down relative to time on Earth. When they return, everyone else has aged more than them. General relativity also opens up the possibility that you can travel back in time, opening up strange paradoxes. Go to the past and you could kill your grandfather, throwing your own existence into confusion. Despite its strange implications, relativity has never failed an experimental test yet.

Einstein's theory of relativity is our most accurate way of describing how matter, energy, space and time interact. It is actually two theories: special and general relativity. Special relativity came first. This said nothing can travel faster than the speed of light. It also showed that the passage of time is different for people travelling at different speeds. According to the theory, if twins are separated by one taking a journey through space at close to the speed of light, they will have significantly different ages when reunited later. Special relativity also spawned the famous equation $E=mc^2$, which describes how matter can be converted into energy, and vice versa. This equation laid the foundations for the atomic bomb and nuclear power.

The later theory, general relativity, overthrew Isaac Newton's concept of gravity. It portrays time as a dimension, just like the three dimensions of space, and combines them all into something called space-time. Anything with energy or mass warps space-time, creating a gravitational field. Such a field has an effect on any matter travelling through it. It even bends passing light rays in a process called gravitational lensing. The first proof of relativity came with the observation of this phenomenon during a solar eclipse in 1919.

RELATED THEORIES
see also
UNIVERSAL GRAVITATION THEORY
page 20

UNIFICATION
page 50

QUANTUM FIELD THEORY
page 46

OCKHAM'S RAZOR
page 142

3-SECOND BIOGRAPHIES
ALBERT EINSTEIN
1879–1955

ISAAC NEWTON
1643–1727

30-SECOND TEXT
Michael Brooks

'Everything is relative'. We often say it but perhaps we do not grasp how true it is – even time, mass and space are connected.

> According to the theory of relativity, the faster you travel through space, the slower you travel through time. If you reach the speed of light, time will stop completely.

THE MICROCOSM

THE MICROCOSM
GLOSSARY

alpha particles Invisible objects released by some radioactive substances. Alpha particles consist of two protons and two neutrons and have an electric charge of +2.

Brownian motion An observed phenomenon in which small objects, such as smoke particles, appear to wobble randomly. The movements are caused by invisible atoms frequently colliding with the visible objects.

dimension A fundamental measure used to describe an object or event. Humans are aware of four dimensions – length, width, height and time – but scientific theories often involve multiple dimensions that are only perceived through mathematics.

electromagnetism The relationship between the forces that push electrons along as currents of electricity and the forces exerted by magnets. It is one of the four forces of nature and relates to electric charges. Opposite charges attract each other; like charges repel. The electromagnetic force gives atoms their internal structure and allows a set of atoms to stick together to form the many substances that make up the Universe.

electron A tiny, negatively charged particle that is found in atoms. Electrons flow through metals to create electric currents. They are also involved in chemical reactions that allow atoms to bond together.

electron microscope A microscope that makes images of very small things using a beam of electrons instead of a beam of light.

element Atoms contain a varying set of smaller particles – protons, electrons and neutrons. Each set gives an atom certain physical and chemical properties. An element is a substance that is made up of just one type of atom. It cannot be broken down into simpler substances. There are about 90 elements found on Earth. Gold, sulphur, oxygen and hydrogen are some examples. Water is a combination of oxygen and hydrogen, and as such is not an element – it can be split into its two constituent parts.

force A phenomenon in which energy is transferred from one object to another. The Universe is held together by just four forces that pull matter together or keep it separated. The four forces are gravity, electromagnetism, the weak nuclear force and the strong nuclear force. Gravity is the weakest but works across the largest distances – it forms stars and keeps planets moving in their orbits. Conversely, the strong nuclear force is the most powerful of the four, but it acts only across distances that are a fraction of the width of an atom.

mechanics The branch of physics that is concerned with forces and motion of matter. In the everyday world, mechanics uses Newton's three laws of motion. However, in quantum mechanics these rules no longer apply, and physicists describe the motion, location and other characteristics of particles in terms of probabilities.

microcosm The little picture – a model of a system that describes its workings in terms of events taking place on the smallest of scales.

neutron A subatomic particle found in the nucleus of every atom.

nucleus The tightly packed centre of an atom containing particles called protons and, normally, neutrons, too. The protons make the nucleus positively charged, which attracts an equal number of electrons to move around it, thus completing the atom. Almost all the mass of an atom is contained within the nucleus.

photons Packets of energy used to transfer electromagnetic forces. Radiation, such as light, heat and X rays, are waves of photons. The name we give to radiation depends on the amount of energy in its photons. Radio waves contain low-energy photons, while X rays and gamma rays carry the highest energies. Light and heat are somewhere in the middle of the range.

probability A measure of chance.

proton A positively charged subatomic particle that is found in the nucleus of atoms.

quantum A unit that cannot be subdivided any further. Energy exists in quanta.

radioactive A property of certain atoms that are too unstable to stay whole. They break apart, or decay, emitting radiation as small, high-speed particles.

semiconductor A substance that can be made to conduct electricity or insulate against it. Semiconductors are used as tiny switches in computers and electronics.

strong nuclear force The force that holds the protons and neutrons together in a nucleus.

teleportation Moving a solid object by breaking it apart and rebuilding it from new atoms somewhere else.

weak nuclear force The force involved in ejecting certain particles from a nucleus during radioactive decay.

ATOMIC THEORY

the 30-second theory

RELATED THEORIES
see also
QUANTUM MECHANICS
page 38

THE UNCERTAINTY
PRINCIPLE
page 40

QUANTUM FIELD THEORY
page 46

3-SECOND BIOGRAPHIES
DEMOCRITUS
c. 460–370 BC

JOHN DALTON
1766–1844

ALBERT EINSTEIN
1879–1955

ERNEST RUTHERFORD
1871–1937

30-SECOND TEXT
Jim Al-Khalili

3-SECOND THRASH
Deep down, everything in the Universe is composed of the same set of building blocks.

3-MINUTE THOUGHT
Today, atomic theory is much more than just a theory; it is an undeniable fact. Not only can we see individual atoms using electron microscopes, we can even trap them and move them around using lasers. So, when we talk about atomic theory today, we do not mean the theory that states that everything is made of atoms, but rather the theory that describes how atoms behave and interact – the realm of quantum mechanics.

The original atomic theory was proposed in the 5th century BC by the Greek philosopher Democritus, who speculated that everything in the world is ultimately composed of combinations of small, hard and indivisible particles. He called the particles atoms, and suggested that they came in various shapes and sizes but were all made of the same basic material. The modern scientific theory of matter states that the great variety of substances we see in the Universe are made from combinations of different chemical elements. These elements do indeed consist of trillions of identical sub-units, or atoms. The internal structure of an atom is specific to each element and gives that element its particular properties and characteristics. Thus, a hydrogen atom is constructed differently from an atom of gold. Modern atomic theory kicked off at the beginning of the 19th century with the work of the English chemist John Dalton. However, it was not until 1905 that Einstein proved the existence of atoms mathematically in his famous paper on Brownian motion. A few years later Ernest Rutherford was the first to look inside atoms in an experiment in which he bombarded a thin sheet of gold with alpha particles. He discovered that every atom consists of a tiny, positively charged nucleus surrounded by empty space in which even tinier, negative electrons orbit around the nucleus.

Everything you have seen, can see and will ever see is made from a jumble of atoms – even you.

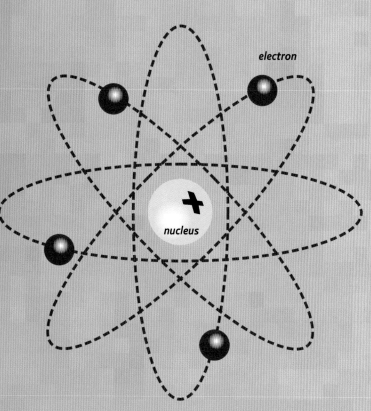

electron

nucleus

> It is impossible to visualize what the inside of an atom looks like, but we normally represent it as a small, central nucleus surrounded by whirling electrons.

QUANTUM MECHANICS

the 30-second theory

RELATED THEORIES
see also
THE UNCERTAINTY
PRINCIPLE
page 40

SCHRÖDINGER'S CAT
page 42

QUANTUM FIELD THEORY
page 46

3-SECOND THRASH
As one of the founding
fathers of quantum
theory, Niels Bohr once
said, 'If you are not
astonished by quantum
mechanics then you
haven't understood it!'

3-MINUTE THOUGHT
Quantum mechanics
is probably the single
most important theory
in physics. Despite the
difficulty we have in
understanding what it
all means, we have it to
thank for almost all of
modern technology. It
describes how atoms stick
together into molecules,
how semiconductors
work, and explains how a
laser functions. Without
quantum mechanics we
would have no computers,
MP3 players, mobile
phones, or life-saving
medical treatments ... and
so much more.

This is the weird yet incredibly
powerful theory of the subatomic world in
which everyday concepts to do with forces
and motion no longer apply in the same way.
Instead, we need a new kind of mechanics
based on what are called 'quantum' rules.
This idea was first developed in the early 20th
century by German physicist Max Planck, who
proposed that energy comes in tiny lumps called
'quanta'. The theory was extended by Albert
Einstein, Niels Bohr, Paul Dirac and Werner
Heisenberg, among several others, in the 1920s.

However, despite its tremendous success,
quantum mechanics remains shrouded in
mystery because, uniquely among scientific
theories, no one really knows how or why it
works. It makes certain predictions about the
microscopic world that go completely against
our common sense. For instance, it explains
how an atom can exist in more than one place
at the same time until we check to see what it
is up to. It also says that an electron can spin
both clockwise and anticlockwise at the same
time until we measure it. These, and many more
strange properties, are not created by problems
with the theory but are simply – or not so
simply, depending on your point of view –
how nature behaves down at this scale.

3-SECOND BIOGRAPHIES
MAX PLANCK
1858–1947

ALBERT EINSTEIN
1879–1955

NIELS BOHR
1885–1962

WERNER HEISENBERG
1901–1976

PAUL DIRAC
1902–1984

30-SECOND TEXT
Jim Al-Khalili

*Quantum mechanics
does not give simple
answers – for instance,
predicting that the
same subatomic particle
can exist in a host of
different states at the
same time.*

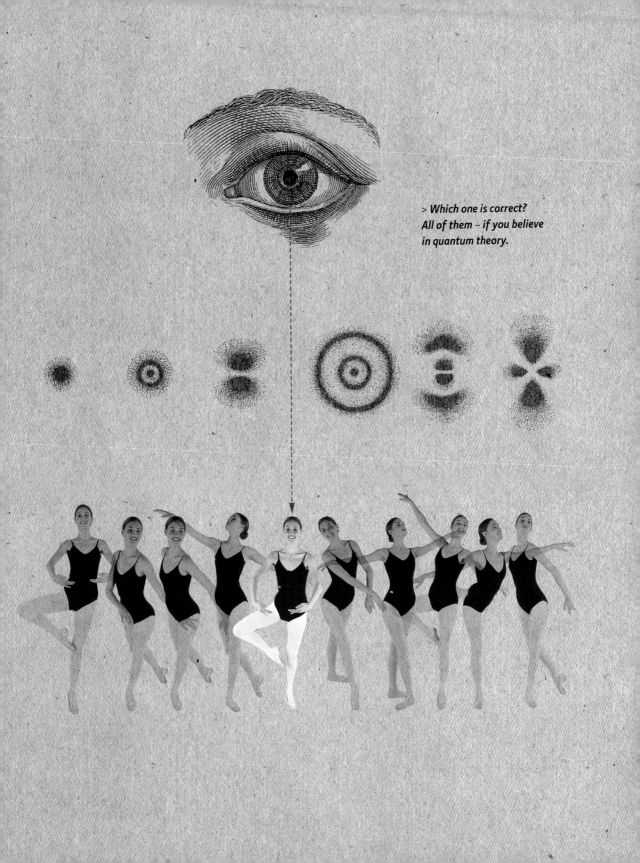

> Which one is correct?
All of them – if you believe
in quantum theory.

THE UNCERTAINTY PRINCIPLE

the 30-second theory

3-SECOND THRASH
The harder you try to pin down a subatomic particle, the faster and more frantically it zips around and tries to escape.

3-MINUTE THOUGHT
The uncertainty principle is often mistakenly taken to mean that the instruments we use to probe the subatomic world are clumsy and inaccurate. In fact, it is really a very powerful statement about the way nature behaves at these tiny scales, and it has many important consequences. For instance, without it the Sun would not shine, since it is the reason hydrogen nuclei are able to fuse together to produce light and heat.

The Heisenberg uncertainty

principle is a statement about the way quantum objects, such as atoms and the smaller particles inside atoms, behave. It was developed in 1927 by Werner Heisenberg, and so bears his name. The principle states that we can never know exactly where an electron, say, is located, while at the same time knowing exactly how fast it is moving. Either property – its speed or position – can be measured to infinite accuracy in principle, provided we sacrifice any knowledge of the other. This is not a shortcoming of our understanding of the workings of nature, nor is it due to the sheer minuteness of an electron, but is simply the way electrons are. In fact, it has nothing really to do with us at all. The electron itself does not have a well-determined position and speed. The best we can do is to identify a region in which the electron is likely to be moving.

Another way of stating the uncertainty principle is in terms of energy and time. We can measure the exact energy of a particle, provided we do not care about when it has this energy. Conversely, if we fix the time of measurement exactly, then we give up any hope of finding out how much energy it has.

RELATED THEORIES
see also
ATOMIC THEORY
page 36

QUANTUM MECHANICS
page 38

SCHRÖDINGER'S CAT
page 42

PARALLEL WORLDS
page 128

3-SECOND BIOGRAPHY
WERNER HEISENBERG
1901–1976

30-SECOND TEXT
Jim Al-Khalili

According to the uncertainty principle, there's only so much you can learn by measuring subatomic particles.

> Where are you, where
have you been and
where are you going?
You'll never truly know.

SCHRÖDINGER'S CAT

the 30-second theory

In the mid-1930s the Austrian physicist Erwin Schrödinger proposed a thought experiment to highlight how crazy quantum mechanics was. He suggested taking a box in which we place a cat, some lethal poison and a radioactive source. According to quantum mechanics we cannot say, unless we are checking, whether a radioactive atom has broken apart, or decayed, within a given time, so we must describe it as having both decayed and not decayed at the same time. Only when we check do we force it to be one or the other.

Inside Schrödinger's box, the experiment is designed so that any decayed atom will have spat out a particle that triggers the release of the poison, killing the cat. Since the cat, said Schrödinger, is also made up of atoms (albeit trillions of them) then it too is presumably subject to the laws of quantum mechanics. So, until we open the box to look, we must describe the cat as being both dead and alive at the same time. Only when we open the box do we force everything inside into one or other state.

RELATED THEORIES
see also
THE UNCERTAINTY
PRINCIPLE
page 40

QUANTUM ENTANGLEMENT
page 48

PARALLEL WORLDS
page 128

3-SECOND BIOGRAPHY
ERWIN SCHRÖDINGER
1887–1961

30-SECOND TEXT
Jim Al-Khalili

3-SECOND THRASH
Since atoms can do two things at once, and a cat is made up of atoms, it can be both dead and alive at the same time.

3-MINUTE THOUGHT
Schrödinger believed there was a flaw in quantum mechanics, since we never see cats simultaneously dead and alive. One argument states that quantum mechanics says nothing about what the cat feels like before we check. It only allows us to calculate the probability that we will find the poor puss dead or alive once we do take a look. Or maybe the whole Universe splits into two when we open the box. In one the cat is alive, in the other it is dead.

Schrödinger used his cat-in-a-box thought experiment in a discussion with Albert Einstein – but it also helps to describe how quantum physics works to the rest of us.

> Ingredient: one cat.
Place in a closed box.

> Next, add a flask of
cat-killing poison
triggered by a single
particle of radiation.
Carefully introduce a
radioactive metal with
a 50:50 chance of firing
off particles.

> Resist the urge to look into
the box. You now have two
half-cats in there; one is
alive, but the other is dead.

1918
Born, New York City,
New York

1939
Graduates from
Massachusetts Institute
of Technology (MIT)

1943
Joins Manhattan Project
to develop atomic bomb

1950
Becomes professor of
physics at Caltech in
Los Angeles, California

1965
Wins Nobel Prize
for Physics

1986
Appointed to the
committee investigating
the explosion of the
space shuttle *Challenger*

1988
Dies, Los Angeles

RICHARD FEYNMAN

Richard Feynman and his many collaborators reconfigured the way we understand quantum physics – the world of impossibly small particles that make up the fabric of the Universe. As with all brilliant scientists, Feynman was a free-thinking maverick. He is known not only for his Nobel Prize-winning breakthroughs, but also for his exhilarating bongo playing!

Richard Feynman was born in New York City in 1918. He was a first-rate student, and even his undergraduate work at MIT on the forces acting inside molecules was noticed by physicists around the world.

Feynman moved to Princeton University in 1941, where he and John A. Wheeler (whose other work also introduced the terms *black hole* and *wormhole*) developed a new theory of 'quantum electrodynamics' – a way of describing electromagnetic fields in terms of the movements of particles. Before that, physicists had visualized them purely as waves.

While at Princeton, Feynman was enrolled in early research into atomic weapons. In 1943 he went to the Los Alamos National Laboratory in New Mexico as the youngest member of the Manhattan Project. There he helped to calculate the explosive power of atom bombs, and set up a primitive computing system to analyze the huge amounts of data involved in the project.

In 1950 Feynman become a professor at the California Institute of Technology (Caltech). While there, he worked with Murray Gell-Man to describe the weak force found inside atoms. This work explained what was happening when radioactive atoms broke apart, or decayed.

Feynman's lectures at Caltech inspired a new generation of particle physicists, and his books on many subjects captivated the general public. Feynman remained at Caltech until his death in 1988.

QUANTUM FIELD THEORY

the 30-second theory

A 'field' in physics is a region of space in which there is some physical influence on objects. Examples include gravitational and magnetic fields. A field theory therefore describes how these fields behave, and how objects interact with the fields they are in.

One of the founders of quantum mechanics, Paul Dirac, published several papers in the late 1920s showing how quantum theory could be combined with James Clerk Maxwell's field theory of electromagnetism, as well as with Einstein's special theory of relativity. What he produced was the first 'quantized' field theory, which described how electrons and photons, the particles of light, interact with each other.

After a promising start, quantum field theory had a rough time during the 1930s and 1940s when it was plagued by mathematical difficulties. These were finally sorted out when, in 1949, several physicists, including the great Richard Feynman, produced quantum electrodynamics, or QED, for short. Later, this theory was used to combine the electromagnetic force with another of the four forces of nature – the weak nuclear force. This development became known as the electroweak theory. A separate quantum field theory, known as quantum chromodynamics, has also been developed to describe the strong nuclear force. Only the fourth and final force of nature – gravity – has, thus far, resisted attempts to be quantized.

RELATED THEORIES
see also
ELECTROMAGNETISM
page 28

ATOMIC THEORY
page 36

QUANTUM MECHANICS
page 38

THE UNCERTAINTY
PRINCIPLE
page 40

3-SECOND BIOGRAPHIES
PAUL DIRAC
1902–1984

JAMES CLERK MAXWELL
1831–1879

RICHARD FEYNMAN
1918–1988

30-SECOND TEXT
Jim Al-Khalili

Quantum field theory applies the bizarre laws of quantum physics not just to solid particles, but also to the fields describing the fundamental forces of nature.

3-SECOND THRASH
This subatomic theory is so accurate, it is like knowing the distance between London and New York to within a hair's thickness.

3-MINUTE THOUGHT
Everything in the world can be fundamentally explained with quantum field theory: everything is made from atoms, which cling together through the interactions of their electrons. These interactions are due to the electromagnetic force acting between them, which in turn is no more than the exchange of photons. Thus, it is fair to say that quantum field theory underpins most of physics, the whole of chemistry and hence all of biology, too.

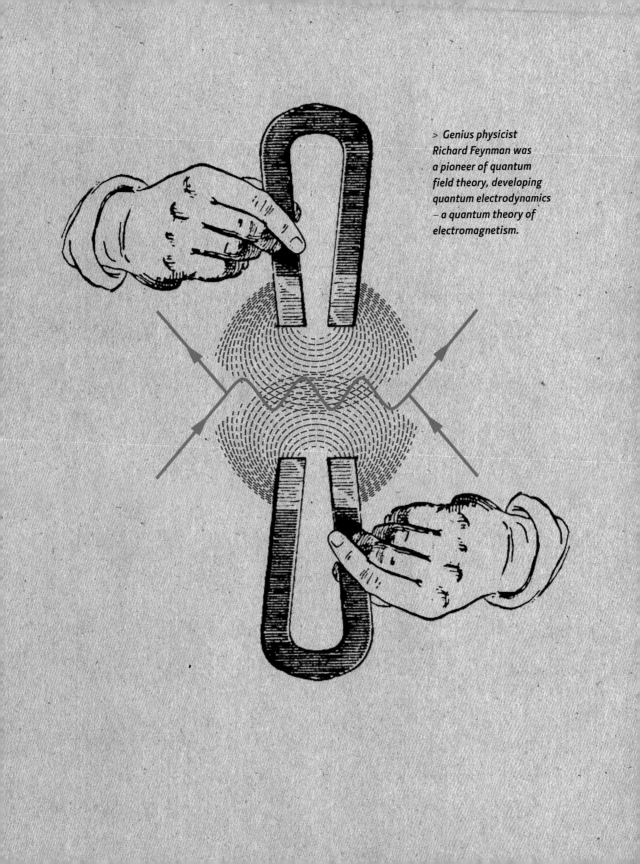

> *Genius physicist Richard Feynman was a pioneer of quantum field theory, developing quantum electrodynamics – a quantum theory of electromagnetism.*

QUANTUM ENTANGLEMENT

the 30-second theory

Not an easy one this! When any two quantum objects, such as electrons or photons, come into contact with each other, their quantum states (the mathematical information describing their properties) combine, or become entangled. Thereafter, their fates remain intertwined, however far apart they move in the future. This bit is not so strange, perhaps, since it is easy enough to believe that, having a shared past means two entities will have affected each other's properties in some way at the time of their interaction. The effect of this interaction can still be seen when we check the particles afterward.

However, entanglement becomes much stranger than that! In the quantum world, entities can exhibit two or more conflicting characteristics simultaneously, such as spinning in opposite directions at the same time. This is called 'superposition'. Now, if a photon, say, is entangled with another it can 'infect' it with its superposition so that they are both in superpositions. However, once we look at one of them, this constitutes a measurement, and we force the photon to decide which way it is spinning. But, because it is entangled with its distant partner, we also force the other photon to make the same choice. This happens instantaneously, even if the two photons are now millions of miles apart.

Subatomic particles appear to be linked with each other, even when divided by impossible distances. Will we ever untangle quantum entanglement?

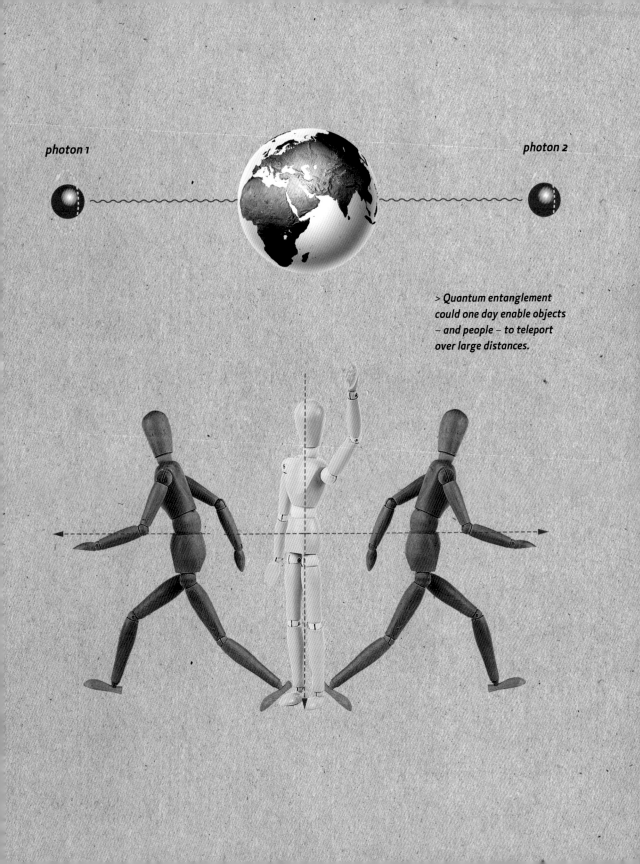

photon 1

photon 2

> Quantum entanglement could one day enable objects – and people – to teleport over large distances.

UNIFICATION

the 30-second theory

This is the attempt to describe

the workings of all four fundamental forces of nature, and the relationships between all elementary particles in a single theoretical framework. In physics, forces can be described by fields that mediate, or carry, the interactions between particles. These are known as field theories. For instance, in 1915 Albert Einstein developed general relativity, a field theory of the force of gravity. At subatomic distances, fields are described by quantum field theories, which apply the ideas of quantum mechanics to the fundamental fields associated with the other three forces: the electromagnetic force and the strong and weak nuclear forces.

The aim of researchers now is to discover whether quantum chromodynamics – the field theory of the strong nuclear force – can be unified with the electroweak theory that describes the electromagnetic and weak forces. The result would be the so-called grand unified theory, or GUT. However, a successful GUT will still not include the force of gravity. The problem is that physicists still do not know how to formulate a workable quantum field theory of Einstein's theory of gravity. One possible candidate for such a 'theory of everything' is called string theory, but we are a long way from knowing whether it is right or not.

RELATED THEORIES
see also
UNIVERSAL GRAVITATION
THEORY
page 20

THEORY OF RELATIVITY
page 30

QUANTUM ENTANGLEMENT
page 48

3-SECOND BIOGRAPHY
ALBERT EINSTEIN
1879–1955

30-SECOND TEXT
Jim Al-Khalili

3-SECOND THRASH
The strong desire among theoretical physicists to find a single theory of everything – with an equation that can fit on their T-shirts.

3-MINUTE THOUGHT
Physicists have been working on unification for almost 100 years. Einstein spent the last 30 years of his life trying to unify gravity with electromagnetism, even before the two nuclear forces were discovered, but without success. What we do know is that in order to unify forces it seems we need a theory with more than four dimensions. This is why string theory, one of the leading contenders for unification, requires ten dimensions – nine of space plus one of time.

Everything in the Universe is connected – easy enough to say, but it's proving difficult to figure out how.

> The greatest minds in history have tried – and failed – to come up with a 'theory of everything'.

HUMAN EVOLUTION

abstract Relating to something imaginary or intellectual. The subject cannot be directly touched or experienced in any other physical way, but humans are capable of sharing abstract notions using speech.

altruism The opposite of selfishness. Human societies encourage altruistic behaviour, such as charity work, because it makes life better for everyone in the long run. Similarly, animal relationships also appear to involve altruism. Meerkats take it in turns to keep a look out for danger – and warn their friends of attack; wolves babysit for each other; and lions hunt in teams and share their food with the whole group. Evolutionary theory tells us that it is the genes that benefit in the long run from animals being unselfish.

amino acid A chemical made up of carbon, hydrogen, oxygen and nitrogen atoms. Amino acids are the building blocks of proteins. The acids are arranged in long chains, which coil up on themselves to make proteins with a well-defined shape. Proteins are often described as the substances in muscles, or meat, but they are the machine tools of life. Each body cell uses hundreds of proteins to manufacture and process the chemicals that keep us alive.

atheist Someone who does not believe in the existence of any gods. This position is slightly different to that of an agnostic, who thinks it is impossible to know whether a god exists or not – and therefore does not bother to be concerned by it.

ecology The study of wildlife communities. An ecologist looks at the way an organism lives in terms of the factors that affect its survival, such as the supply of food, the weather and the activities of all the other animals, plants and various life forms that share its habitat. Ecology can be used to describe the way members of the same species relate to each other as they compete or cooperate to control resources. It also builds up a picture of the relationships between species, creating a model called an ecosystem.

genetics The study of genes. The term gene has two main definitions. The first is a unit of inheritance. This perhaps relates best to the most popular concept of a gene in use today. When people say that they have the gene for red hair, we all understand that they mean they inherited this characteristic from their parents, and will pass it on to their own children. However, this definition does not tell us much about what physical factors are responsible for giving them red hair. The second definition for gene is a strand of DNA. DNA is the complex chemical that carries the blueprints for a living body in code. A gene is a section of this DNA that carries a meaningful chunk of code that can be translated into a working part of a living cell. One of the key roles of geneticists is to identify the relationships between these two varying definitions of what a gene is by determining which bits of DNA translate into measurable characteristics.

Homo The genus, or group, to which human-like animals belong. Today, only one species of human survives, but in the past there have been a handful of others. They include *H. habilis*, which means Handy Man, and *H. erectus*, or Upright Man. *Homo sapiens*, our own scientific name, means Wise Man.

hypothesis A set of ideas that has yet to be tested by experiments. If the tests do not prove the ideas within the hypothesis to be wrong, then the hypothesis is upgraded to a theory – and considered to be true, until another theory succeeds in showing it to be false. The job of scientists is to use these theories to formulate new hypotheses and seek out more truth.

prebiotic chemicals Relating to before life existed, on Earth at least. Life is based on several classes of chemicals. Some are complex but most, such as sugars and amino acids, are relatively simple. When we find them in the environment today, we may assume that they were made by a life process. However, we also find them in space. It is believed that the first living entities made use of these so-called prebiotic chemicals in a 'primordial soup'. The leading theory is that they arose through purely chemical processes, but it has been suggested they might have arrived from space.

psychology The scientific study of the human mind. It should not be confused with the more subjective field of psychoanalysis.

spore A small, seed-like structure that is capable of growing into a full-sized living body, or multiplying into a colony of single-celled organisms, such as bacteria. Certain types of bacteria have a spore phase in which they become enclosed in a cyst inside a hard case that protects the cell inside from all but the most powerful sterilization techniques. Other life that produce spores include fungi, ferns and some parasitic animals, such as tapeworms, which produce eggs as toughened spore-like cysts.

territoriality The practice by animals of controlling areas or home ranges. Territories provide animals with both food and locations for their dens.

PANSPERMIA

the 30-second theory

The Swede Svante Arrhenius

suggested a century ago that life in the form of spores could survive in space and be spread from one planetary system to another. According to Arrhenius, spores escape by random movement from the atmosphere of a planet, and are spread throughout interstellar space by the weak but persistent radiation pressure exerted by starlight. In a variation on this theme, others have suggested that the spores might be spread deliberately by intelligent beings. This theory is called 'directed panspermia'.

The modern version of the panspermia idea starts from observations of prebiotic chemicals in interstellar clouds. It seems certain that some of these raw materials, such as amino acids, fell on to the young Earth and kick-started life. Some researchers, notably the late Fred Hoyle and Chandra Wickramasinghe, have argued that not just complex organic substances, but even complete living organisms, albeit bacteria, might have evolved in space on the surface of dust grains, then been carried down to Earth by an impacting comet. There is also the possibility of 'ballistic panspermia', when rocks from one planet are blasted into space by impacts and travel to another planet. The discovery on Earth of meteorites from Mars' surface means that we might even be descended from Martian bugs.

3-SECOND THRASH
Life on Earth may have been seeded by spores flying in from outer space.

3-MINUTE THOUGHT
The astronomer Thomas Gold suggested in the 1960s that explorers from a space-travelling race might have accidentally left behind contamination after visiting Earth. He imagined them having a picnic and leaving crumbs behind. Carl Sagan pointed out that, in that case, 'Some microbial resident of a primordial cookie crumb may be the ancestor of us all.'

RELATED THEORIES
see also
NATURAL SELECTION
page 58

THE ANTHROPIC PRINCIPLE
page 122

3-SECOND BIOGRAPHIES
SVANTE ARRHENIUS
1859–1927

FRED HOYLE
1915–2001

CHANDRA
WICKRAMASINGHE
1939–

30-SECOND TEXT
John Gribbin

Did life arrive on Earth from outer space? And have spores from Earth seeded life on other worlds?

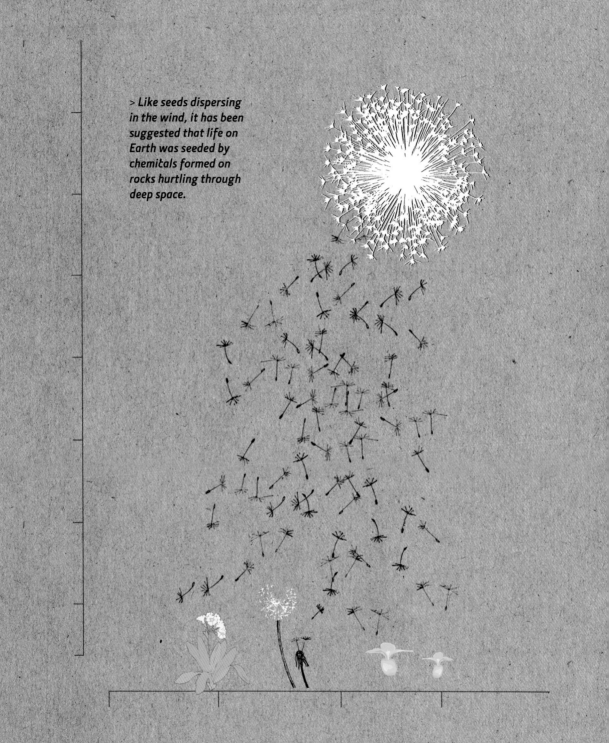

> Like seeds dispersing in the wind, it has been suggested that life on Earth was seeded by chemicals formed on rocks hurtling through deep space.

NATURAL SELECTION

the 30-second theory

3-SECOND THRASH
Living things are shaped
by the environment they
live in, which is why
dolphins look like sharks
but not camels.

3-MINUTE THOUGHT
The theory of natural
selection was thought up
independently by both
Charles Darwin and Alfred
Wallace, but Darwin
published his ideas first in
1859. It revolutionized not
only biology, but also much
of human thinking, as it
undermined traditional
creation stories and,
thus, the existence of
God. Today, evolutionary
biology is studied at the
level of genes. Every form,
function and behaviour
of a living body is now
described in terms of
'selfish genes'. The only
purpose of life is to make
ever more copies of genes.

A person who is suffering from a sore throat or earache is infected with bacteria, which are spreading through those parts of the body. Antibiotics are used to treat the illness. A small dose of the drug will disrupt the way most of the bacteria reproduce, so the infection dies away almost completely. However, a few days later the infection may return, as bacteria that are resistant to the antibiotic take the place of those removed by the treatment. The change from a drug-susceptible to a drug-resistant bacterial population is a particularly rapid example of natural selection in action. Drug-resistance is passed on to the next generation of bacteria, and eventually all members of the population share this trait.

This process occurs in any entities that can pass on characteristics to their offspring as they reproduce. Crucially, however, mistakes in the process ensure that each individual offspring is unique, if only in tiny ways. In the patient with the infection, the drug-resistant bacteria reproduce more successfully than the bacteria that are affected by the drug.

Evolutionary change normally takes much longer, but the same process that alters a group of bacteria also explains the evolution of life on Earth over 3,500 million years. Natural selection ensures that life adapts to survive in different habitats, and will continue to evolve as the prevailing conditions change.

RELATED THEORIES
see also
THE SELFISH GENE
page 60

MEMETICS
page 144

3-SECOND BIOGRAPHIES
CHARLES DARWIN
1809–1882

ALFRED WALLACE
1823–1913

30-SECOND TEXT
Mark Ridley

*Natural selection says
that species evolve to
overcome the challenges
presented by their
environment.*

> The theory of evolution by natural selection is a knock-out punch for religion. It's able to explain the emergence of complex organisms that were previously thought to require a Creator.

THE SELFISH GENE

the 30-second theory

The attributes of living things

appear to benefit the individuals that possess them. When a cat eats a mouse, the cat deploys sensory systems (eyes, whiskers, etc.), muscles, claws and digestive systems, all to enable it to survive for another day. Those attributes evolved by natural selection. Natural selection therefore seems to produce adaptations that benefit individual organisms. However, these same organisms are sometimes altruistic – one individual sacrifices itself for the benefit of another. Altruism seemed to fly in the face of natural selection, and it was often said that the sacrificial behaviour was 'for the good of the species'. However, no mechanisms could be found to explain natural selection at the level of whole organisms. Altruistic behaviour makes more sense when we look how it benefits the genes. Close relatives, such as two sisters, share at least some of the same genes. One sister may sacrifice herself to ensure the survival of the other. The dead sister cannot pass on her genes herself. However, the genes she shares with her sister will survive her death, so it still benefits her 'selfish genes' to die protecting her sister. Of course, sacrificing one's life is an extreme form of altruism. Most animals can enhance their kin's survival by warning them of danger and sharing food. However, in some ecologies, suicidal defence is the norm, such as when honey bees sting threats to the hive. They will die in the attack, but their sisters will survive.

RELATED THEORIES
see also
NATURAL SELECTION
page 58

OCKHAM'S RAZOR
page 142

MEMETICS
page 144

3-SECOND BIOGRAPHY
RICHARD DAWKINS
1941–

30-SECOND TEXT
Mark Ridley

3-SECOND THRASH
We all exist merely to carry genes and make as many copies of them as possible.

3-MINUTE THOUGHT
The theory of selfish genes was popularized by Richard Dawkins in a book of the same name in 1976. Dawkins has since become a controversial figure as a vocal atheist. He includes the theory of selfish genes in critiques of theology. He has also spent many years clarifying the relationship between genes and the characteristics they express. Contrary to a widely held belief, the latter is not wholly determined by the former. Instead, the genetic component must interact with the environment to produce the final characteristic, the oft-debated balance between nature and nurture.

The selfish-gene theory dates from the 1960s, but the novelist Samuel Butler managed to sum it all up in the 19th-century when he said, 'Which came first, the chicken or the egg? The chicken is merely the egg's way of making another egg.'

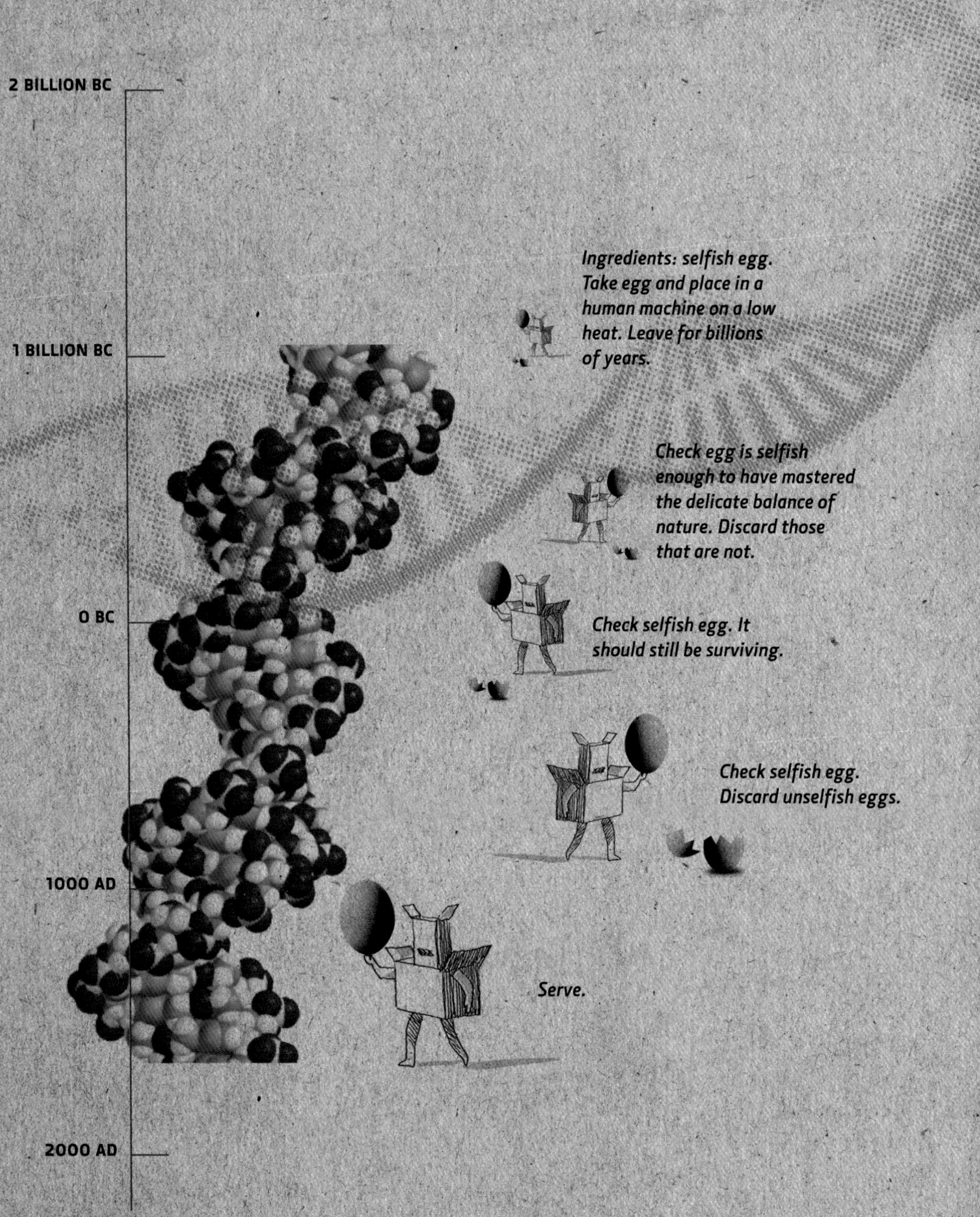

2 BILLION BC

1 BILLION BC

Ingredients: selfish egg.
Take egg and place in a
human machine on a low
heat. Leave for billions
of years.

Check egg is selfish
enough to have mastered
the delicate balance of
nature. Discard those
that are not.

0 BC

Check selfish egg. It
should still be surviving.

Check selfish egg.
Discard unselfish eggs.

1000 AD

Serve.

2000 AD

1809
Born, Shrewsbury, England

1825
Begins medical studies at University of Edinburgh

1831
Sails with HMS *Beagle* to South America and the Pacific

1839
Elected a Fellow of the Royal Society

1842–44
Produces essays on the subject of natural selection

1859
Publishes *On the Origin of Species*

1864
Awarded the Copley Medal, the highest accolade of the Royal Society

1882
Dies, Downe, England

CHARLES DARWIN

Professional scientists have their own favourites and heroes who have been an inspiration to them, but perhaps Charles Darwin's theory of evolution by natural selection has done more than any other scientific work to change the way ordinary people view their place in the world.

In 1809, Charles Darwin was born into an illustrious family. His maternal grandfather was Josiah Wedgwood who had made his fortune mass-producing fine china. Darwin's paternal grandfather was Erasmus Darwin, a hard-drinking doctor and poet, who had himself written a book about evolution in the 1790s. Erasmus proposed that animals changed their forms under the direct influence of the environment. This concept was later expanded by Jean-Baptiste Lamarck, and is now better remembered as Lamarckism.

Charles Darwin was sent to study medicine in Edinburgh at the age of 16. It was there that Darwin was enthused by Lamarckism, and the work of geographer Alexander von Humboldt and geologist Charles Lyell, who were suggesting that Earth was much older than previously thought.

In 1831, Darwin paid to join HMS *Beagle* on a voyage to survey the coasts of South America. What Darwin saw during his 18 months aboard led him to formulate his theory of evolution by natural selection. However, he chose to keep it secret until the late 1850s. While working on a book on the subject in 1858, Darwin received a letter from fellow naturalist Alfred Wallace. Wallace had come up with a similar theory while working in Indonesia. This spurred Darwin to publish *On the Origin of Species* the following year. The work rocked the scientific community. Darwin shrank from public debate on the matter. He left that to others, and retired to Downe House in Kent. There, he wrote several more books, most notably on breeding strategies and the function of emotions. Charles Darwin died in 1882.

LAMARCKISM

the 30-second theory

3-SECOND THRASH
Do body-builders produce
muscly children?

3-MINUTE THOUGHT
For most of the 20th
century, Lamarckism was
a byword for bad science.
Lamarck is one of the
unlucky few people who
have the misfortune of
being remembered for
getting it spectacularly
wrong. However, perhaps
we should not speak too
soon. New research into
the process of cell division
shows that non-genetic
factors are passed from
the parent cells to the
offspring along with
genes. Are these factors
acquired during the life of
the cell? Do they influence
the new cells in some
way? Have we found a
mechanism for Lamarckism
after all these years?

Every living creature is born with some attributes, and acquires others during its lifetime. Acquired characteristics include deformities, such as the pockmarks of disease, and the scars of healed wounds; body-building developments, such as enlarged muscles due to exercise; and learned skills, such as an ability to read. It has been widely believed, since the beginning of recorded human thought, that some of these acquired attributes are inherited by the next generation. The oft-cited example of this concept is that well-muscled blacksmiths will tend to produce strong children.

This idea can be found in the writings of Plato, in ancient Greece. In the 19th century the French biologist Jean-Baptiste Lamarck incorporated the inheritance of acquired characteristics in his pre-Darwinian theory of 'evolution'. A famous example of Lamarck's argument concerns the giraffe. Generations of giraffes may have reached up into trees for food, slightly stretching their necks. If the acquired stretched neck of each individual was inherited by its offspring, over time the giraffes would evolve longer necks. Biologists continued to believe in the inheritance of acquired characteristics through the 19th century. It is something of a historical accident that Lamarck's name happens to refer to the inheritance of acquired characteristics, but, nevertheless, the name stuck.

RELATED THEORIES
see also
NATURAL SELECTION
page 58

3-SECOND BIOGRAPHIES
PLATO
428/427–348/347 BC

JEAN-BAPTISTE LAMARCK
1744–1829

30-SECOND TEXT
Mark Ridley

If there isn't enough grass for everyone, why not try some leaves? That is how giraffes got their long necks. It just didn't happen the way Lamarck suggested.

> *Stretch for success. The middle giraffe is going to get very hungry if it doesn't try harder.*

growth of neck

1st generation   2nd generation   3rd generation

OUT OF AFRICA

the 30-second theory

The ancestors of humans split

from the other great apes about six million years ago. Between then and about two million years ago, all our ancestors lived in Africa. Then some proto-humans, called *Homo erectus*, spread from Africa and colonized Europe and Asia as well. Until about 30,000 years ago, *H. erectus* and its descendants – known as Neanderthals in Europe – occupied all three of these continents. Fossilized remains tells us that modern human beings differ from these proto-humans, particularly in brain shape. The question is, did modern humans evolve from the proto-human species living across Africa and Eurasia – an idea called 'the multiregional hypothesis'? Or did we all evolve from a single ancestor in Africa, who spread across the globe in a second wave of migration that wiped out the indigenous proto-humans? The fossil evidence has been indecisive. Then, in the 1980s, came the genetic evidence. If the multiregional hypothesis is right, the genes of all modern humans should have a common ancestor about 2 million years ago. In fact, the amount of difference between our genes is very small. The degree of similarity implies that we share a common ancestor that is much younger than two million years old – more like 100,000 years. In that time evolution has diversified our DNA only slightly. With such a recent common ancestor, all the evidence now points to modern humans being related to a single African ancestor.

RELATED THEORIES
see also
SOCIOBIOLOGY
page 68

THE ORIGIN OF LANGUAGE
page 70

RARE EARTH HYPOTHESIS
page 110

30-SECOND TEXT
Mark Ridley

3-SECOND THRASH
Every human being today is descended from an ancestor who lived in Africa about 100,000 years ago.

3-MINUTE THOUGHT
The 'out of Africa' hypothesis implies that Neanderthals are a distant side branch of the modern European family tree. Even though Neanderthals occupied Europe from about 200,000 to 30,000 years ago, none of their genes appear to survive in modern European populations. Amazingly, genes have been extracted from some deep-frozen Neanderthal fossils. The Neanderthal genes are completely different from any genes found in humans today. Once again, genetics gives a decisive answer to a long-debated question.

Genetics proves that all 6.7 billion human beings alive today are descended from just a few thousand early Africans.

> *Wherever you are
from, we were all
Africans originally.*

SOCIOBIOLOGY

the 30-second theory

3-SECOND THRASH
From religious extremism to why men won't ask for directions – it's all down to biology.

3-MINUTE THOUGHT
Despite the term *sociobiology* being associated with attempts by biologists to explain away the intricacies of human society, biologists have continued to research non-human social behaviour. They often avoid Wilson's term and call themselves behavioural ecologists, instead. Likewise, people continue to discuss biological, and particularly evolutionary, explanations for human social behaviour. They often call their subject evolutionary psychology.

The word 'sociobiology' entered the popular consciousness following the publication of a book of that name in 1975 by the Harvard biologist E. O. Wilson. Wilson defined sociobiology as the biological study of all social behaviour. Wilson was writing about an emerging area of biological research that sought the answers to many questions: why are males and females different? What is the function of mating systems – why are some species monogamous, but others are decidedly not? Why are animal societies structured the way they are? For example, chimpanzee groups consist of male relatives, with adolescent females moving away from home, while in baboons it is young males that leave their families. Sociobiology also tackles the reasons behind cooperation and altruism, territoriality and why some species live in large groups while others have loners.

However, Wilson went one step further. In the last chapter of his long book he suggests some biological explanations for human social behaviour. The chapter provoked opposition from people who thought their political views were threatened. Despite its roots in animal behaviours, 'sociobiology' in popular discussion usually refers to biological explanations for human social behaviour.

RELATED THEORIES
see also
THE ORIGIN OF LANGUAGE
page 70

3-SECOND BIOGRAPHY
E. O. WILSON
1929–

30-SECOND TEXT
Mark Ridley

If you think human society is the pinnacle of sociobiology, think again. Bees, ants and other social insects live in large groups as well, but for very different reasons.

> *The residents of a beehive are all sisters working for their mother, the queen. Their reason for being is to look after their baby sisters – yet more daughters of the queen.*

THE ORIGIN OF LANGUAGE

the 30-second theory

Grammar is the key to the origin of human language. Many non-human species have a rich system of signs that they use for communication, but they lack the subordinate clauses, moods, cases and prepositions that make up grammar. Human language is infinitely expressive and can be used to discuss – and think about – abstract possibilities, as well as give signs and commands. When did grammatical language originate? The main changes required were in the software of the ancestral human brain, and that cannot be studied directly. However, we have two indirect pieces of evidence. One is the time when human brains arose that worked in the same way as ours. This was probably a little more than 100,000 years ago, and the brains belonged to the humans that spread from Africa and gave rise to the talkative ape called *Homo sapiens* – us. Their brains were anatomically identical to ours, and that suggests they shared our linguistic abilities. However, perhaps the key change came later, in how we use our brains, rather than in crude anatomy. In the archaeological record, there is a great explosion in the richness and artistry of tools, cave decorations and other objects around 30,000 years ago. Maybe this reflects the origin of language as we know it – as a tool for sharing our inner thoughts, plans and ideas.

RELATED THEORIES
see also
OUT OF AFRICA
page 66

SOCIOBIOLOGY
page 68

RARE EARTH HYPOTHESIS
page 110

30-SECOND TEXT
Mark Ridley

3-SECOND THRASH
When exactly did humans stop grunting at each other and start to have civilized chats?

3-MINUTE THOUGHT
Another source of evidence has recently become available: genetics. At the dawn of language, changes presumably occurred in the genes that shaped the parts of the brain concerned with communication. Maybe we will be able to identify these genes and date the times of their rapid change. One such gene, called FOXP2, is associated with linguistic ability. It underwent a spurt of evolution around 120,000 years ago. As more such 'language genes' are identified, they could provide the best date yet for when we learned to talk.

Language is what sets humans apart from the beasts. Only we can express the contents of our minds. All we have to do now is think of something to say.

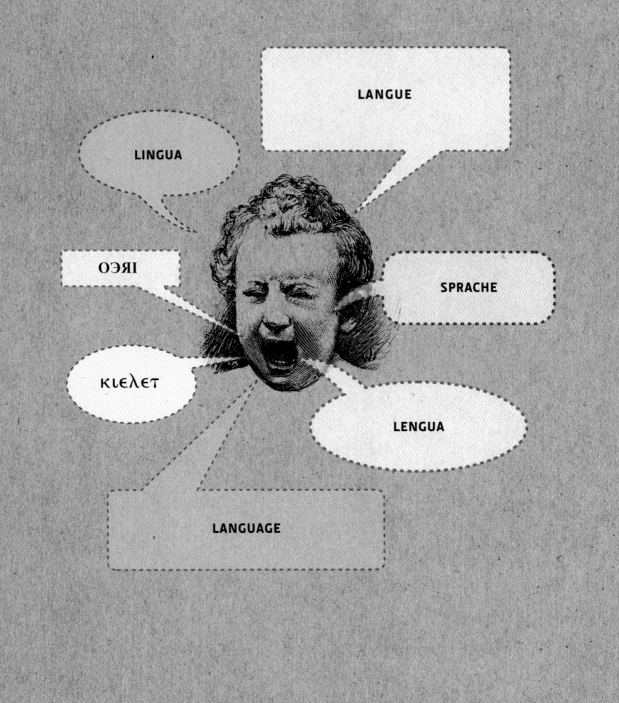

MIND & BODY

artificial intelligence An area of computer science that attempts to deconstruct the processes underlying intelligence and then program them into a computer, so that it can learn and think for itself in the same way as a human being.

conditioning When a particular behaviour becomes associated with a stimulus or signal. The link is reinforced by rewards for the correct behaviour or punishments for incorrect responses. The most famous example of this is Pavlov's experiment with dogs.

determined When a behaviour is completely controlled by a set of rules that are already in place. Instincts could be described as determined behaviours. Most human behaviours are thought not to be determined, but the result of thinking, or cognition.

Freudian slip According to Sigmund Freud a person sometimes reveals his or her true, but repressed, feelings by mistakenly confusing his or her words. Typical Freudian slips relate to personal relationships, such as a man calling his wife 'Mum'.

gene This word has two main definitions, which are quite distinct from each other. The first definition is that a gene is a unit of inheritance. This perhaps relates best to the most popular concept of a gene in use today. When people say that they have the gene for blue eyes, we all understand that they mean that they inherited this characteristic from their parents. However, this definition does not tell us much about what physical factors are responsible for giving them blue eyes. The second definition for gene is a strand of DNA. DNA is the complex chemical that carries the coded blueprints for a living body. A gene is a section of this DNA that carries a meaningful chunk of code that can be translated into a working part of a living cell.

homoeopathy A branch of alternative medicine that attempts to treat disease using very tiny amounts of active ingredients. Treatments are prepared by diluting a highly concentrated substance in water. Proponents of homoeopathy believe that this initial high concentration of ingredients 'activates' the water, giving it medicinal value. Some homoeopathic treatments are so diluted that only a fraction of a batch will actually contain a single molecule of the active ingredient. The rest of the batch will have no active ingredient at all, so you might be treated with nothing but pure water.

hypothesis A set of ideas that seek to explain an observed natural phenomenon but which has yet to be tested by experiments. If the tests do not prove the ideas to be wrong, then the hypothesis is upgraded to a theory – and considered to be true. That is, until another theory succeeds in showing it to be false. The job of scientists is to use these theories to formulate new hypotheses and seek out a better version of the truth.

learning theory A branch of behaviourism that describes how people alter their behaviour patterns and beliefs as a result of their experiences. The mainstream theory is called constructivism, which suggests that people are constantly constructing and modifying their motivations as a result of their current and past experiences, or the experiences of those around them.

linguistics The study of languages. Linguists study syntax (sentence structures), grammar (the language's rules) and pronunciation. This allows them to group languages into families and attempt to figure out how languages have changed over time. This provides an insight into the movement of people around the globe during prehistory.

neuroscience The study of the functioning of the brain and other parts of the nervous system. It focuses on the anatomy and biochemistry of the system and relates these factors to psychology.

neurotic Relating to an emotional disorder, or neurosis. A neurosis condition differs from a psychosis, which relates to a personality disorder and a distorted view of reality. A neurosis might result in a tic – an unconscious physical twitch, usually on the face.

paradigm A grand idea comprising a set of assumptions and concepts that underlies a view of a subject. A 'paradigm shift' occurs when a new discovery alters the way everyone sees the world, for example when Earth was discovered to be spherical, not flat.

psychology The scientific study of the human mind. It should not be confused with the more subjective field of psychoanalysis.

sensory cortex The section of the forebrain that processes information coming from the body's senses – eyes, ears, touch, etc. The information arrives via a network of nerves devoted to carrying sensory signals. Motor responses – that is, movements – are ordered by a separate cortex.

utopian A perfect society in which citizens follow a strict set of rules to ensure peace and justice for all.

the 30-second theory

Rather like those cartoons that

show a person with a devilish incarnation of themselves on one shoulder and a more angelic version on the other, Sigmund Freud's psychoanalytic theory posits that the human mind is torn. Our 'ego' struggles perpetually to balance the demands of its moralizing 'superego' with the desires of the insatiable, animalistic 'id'. Central to psychoanalysis is the idea that much of this conflict bubbles away beneath the level of conscious awareness, just occasionally spilling over as a neurotic tic or Freudian slip.

Sex plays a big part in the theory, too. In particular, Freud believed that a common source of psychic conflict derives from sexual fantasies in childhood. The most infamous example is the Oedipus complex – the idea that children lust after their opposite-sex parent, and jealously wish for the death of their same-sex parent. You may deny you ever had such desires. To do so would serve as a perfect example of repression, one of the many defence mechanisms Freud said we use to curb our unacceptable desires. Such mental manoeuvres save face in the short-term, but could ultimately leave you unhinged and in need of a psychoanalyst.

3-SECOND THRASH
According to Sigmund Freud, we can be driven mad by our attempts to tame the sex-hungry animal within us all.

3-MINUTE THOUGHT
A criticism levelled at psychoanalysis is that its claims are untestable. This would have mattered less had Freud not claimed his work was science. Although psychoanalytic psychotherapy has now fallen out of fashion, many of Freud's ideas are finding support from new research. For example, research at the University of Oregon in the USA has shown that the repression mechanism does exist. The researchers found that deliberately forgotten memories are less likely to be recalled in the future.

RELATED THEORIES
see also
SOCIOBIOLOGY
page 68

BEHAVIOURISM
page 78

COGNITIVE PSYCHOLOGY
page 80

3-SECOND BIOGRAPHY
SIGMUND FREUD
1856–1939

30-SECOND TEXT
Christian Jarrett

The inkblot test is a tool used by psychoanalysts. Patients describe the shapes they see in the blots – and in so doing reveal their repressed emotions. But if all you can see is an inkblot, then perhaps you are in real trouble!

> Look inside yourself.
Who do you see
looking back?

BEHAVIOURISM

the 30-second theory

Behaviourism dominated the
science of psychology for half a century, even
though it eliminated all concepts of mind or
consciousness, and refused to take feelings,
thoughts or desires into account. Instead, it
focused on behaviours that could be measured
in the lab. Behaviourism began in 1913 when
the American psychologist John B. Watson
argued that human learning could be studied
like Ivan Pavlov's famous dogs, which were
conditioned to salivate at the sound of a bell
by pairing it with the arrival of their food.
Appropriate conditioning could be used, he said,
to turn young children into any kind of person
he wanted. B. F. Skinner, another American
psychologist, developed 'operant conditioning',
in which rats or pigeons are rewarded for some
behaviours and punished for others, and so
learn to press levers, run through mazes or
peck at colours. Pigeon-guided missiles were
successfully developed from this idea, but
were never used in anger. He also discovered
that rewards are far more effective than
punishments – a crucial principle for parents
and teachers. Skinner founded his own school
of 'radical behaviourism', arguing that all our
actions are determined, and he speculated on
a utopian society based on the conditioning of
all citizens. Behaviourism is no longer such a
dominant force within experimental psychology,
but the findings of the field's learning theory
are still widely applied in education and therapy.

RELATED THEORIES
see also
SOCIOBIOLOGY
page 68

COGNITIVE PSYCHOLOGY
page 80

3-SECOND BIOGRAPHIES
JOHN B. WATSON
1878–1958

B. F. SKINNER
1904–1990

30-SECOND TEXT
Sue Blackmore

3-SECOND THRASH
Nothing counts in
psychology except
behaviours that can be
measured and tested.
Do not talk about mind or
consciousness – they are
unscientific illusions.

3-MINUTE THOUGHT
By turning psychology into
a science, behaviourism
was an advance on the
psychoanalytic speculations
of Freud that preceded
it. It finally fell out of
favour in the 1960s
for two reasons. First,
things happen inside our
heads that cannot be
measured by behaviourist
experiments – even
rats and pigeons have
emotions, intentions and
mental models of their
environment. Secondly,
much of our intelligence
and personality is
inherited, not learned. So
we need an evolutionary
psychology to understand
human nature.

*Behaviourism offers
an explanation of how
people learn through
their experiences –
both good and bad.*

> Is our behaviour just a response to rewards and punishments? Answer 'Yes' and win a prize!

COGNITIVE PSYCHOLOGY

the 30-second theory

Cognitive psychology treats

human beings as information-processing systems, and studies how we think, perceive, learn and remember. The word *cognitive* comes from the Latin word meaning 'to think'; and the term 'cognitive psychology' was coined by Ulric Neisser in 1967. This was an exciting change from the prevailing behaviourism, which had rejected all research into mental processes. The new cognitive psychology studied inner mental events, and tapped into the rapidly expanding science of artificial intelligence. In this new view, the mind was analogous to software and the brain to the hardware of a biological computer. Human actions, decisions and thoughts were all forms of information processing, depending on input data from the senses. So, for example, studies of the brain's visual system revealed how information entering through the eyes is processed by layers of cells in the eye, the mid-brain and the sensory cortex, and on through to areas of the brain that can recognize objects, or control actions and speech. Cognitive psychology expanded dramatically to become the dominant paradigm in psychology towards the end of the 20th century, and is now part of the interdisciplinary field of cognitive science, which also includes parts of neuroscience, linguistics and philosophy.

RELATED THEORIES
see also
THE PLACEBO EFFECT
page 90

3-SECOND BIOGRAPHY
ULRIC NEISSER
1928–

30-SECOND TEXT
Sue Blackmore

3-SECOND THRASH
The human brain is the hardware and the mind the software of a vast biological computer. Everything we do is information processing.

3-MINUTE THOUGHT
Cognitive psychology was so successful that, at the end of the 20th century, most research psychologists called themselves 'cognitive psychologists'. However, it depends on the idea that the brain works by building complex representations of the world. This emphasis on the brain has now begun to give way to 'enactive' theories that take much more account of the rest of our bodies and of our dynamic interactions with the world around us.

Is it a brain or is it a computer? Cognitive psychologists say there's little difference, and that the brain is an information processor.

> *If the brain is a computer it can be reprogrammed by a cognitive psychologist.*

1856
Born, Freiberg, Moravia

1873
Begins to study medicine at the University of Vienna, Austria

1885
Works at the Salpêtrière Hospital, Paris. with neurologist Jean-Martin Charcot

1900
Publishes *The Interpretation of Dreams*

1902
Becomes Professor of Neuropathology at the University of Vienna

1920
Publishes *Beyond the Pleasure Principle*

1938
Leaves Austria after it is annexed by Nazi Germany

1939
Dies, London, England

SIGMUND FREUD

Sigmund Freud was the first

psychoanalyst. He sought to treat mental illness by focusing on the contents of the mind rather than the workings of the brain. Although many of Freud's theories are now regarded as outdated, his work marked a turning point in human civilization. Freud expanded the intellectual world by adding psychology to religion, politics and economics as subjects used to describe our society. Sigmund Freud was born in Freiberg, a city now within the Czech Republic, but then part of the Austrian Empire.

For a man who wrote so much about the significance of childhood relationships it is fascinating to have an insight into Freud's personal life. Commentators suggest that Freud did not get on with his father. Freud's half-brothers were considerably older, and his closest childhood companion was his nephew John. The love–hate relationship, such as that between the young John and Sigmund, is a central plank of Freud's theories.

Freud trained as doctor and specialized in the brain. In 1885, Freud spent several months in Paris, where he met Jean-Martin Charcot. It was Charcot who introduced Freud to the idea that mentally ill patients might have a problem with their mind, rather than brain function. That same year Freud married Martha Bernays. Soon after he began a close friendship with Wilhelm Fliess, a Prussian doctor. Some have questioned the nature of this relationship, pointing to Freud's repeated inclusion of bisexuality in his theories.

Freud's work in psychoanalysis began in earnest in the 1890s and 1900s. It was at this time that he came up with several, now well-known, concepts, such as Freudian slips, word association, the superego, the pleasure principle and penis envy. In his later years, Freud turned his attention to religion and social taboos. He decided to leave Vienna when Austria came under Nazi control in 1938 and died the following year in London.

GENETIC MEDICINE

the 30-second theory

Genes govern the most basic functions of life. When they go wrong we can end up with diseases ranging from Alzheimer's to cancer, and pass on inherited diseases to our offspring. Genetic medicine seeks to find better treatments, if not outright cures, by focusing on the genes responsible for ill health. In principle, it sounds pretty simple: find the dodgy genes responsible for the disease, whip them out, and stick in fresh copies. For quite a while, that is how genetic medicine was portrayed. The reality has, inevitably, proved to be far more complicated. While some inherited diseases, such as cystic fibrosis, are the result of a single faulty gene, most – including cancers – are the result of complex interactions between a host of genes. Repairing even one defective gene has proved incredibly difficult, and to date no one has been cured of a single common genetic disease. Genetic medicine has proved more successful in the development of drugs that fix gene-related defects. For example, the drug Herceptin® was found by identifying genes involved in certain forms of breast cancer. Even so, the drugs have very limited effectiveness, and are another example of genetic medicine falling short of the hype.

RELATED THEORIES
see also
THE SELFISH GENE
page 60

EVIDENCE-BASED MEDICINE
page 88

3-SECOND BIOGRAPHY
VICTOR MCKUSICK
1921–2008

30-SECOND TEXT
Robert Matthews

3-SECOND THRASH
Our genes play a key role in our health, but turning that information into cures for diseases is far harder than it seems.

3-MINUTE THOUGHT
The gene-centred view of the processes of life increasingly looks a bit naive, prompting drug researchers to adopt a more sophisticated view of living systems, of which genes are just a part. Known as 'systems biology', this view aims to understand disease in terms of interactions between genes, cells, organs and the whole organism. While far more complex, it is already paying off, in the form of drugs with far fewer side effects.

Gene therapy might one day be a panacea – but first geneticists need to understand exactly how a strand of DNA builds a human body. Once we know that, we will be able to see when it goes wrong – and fix it.

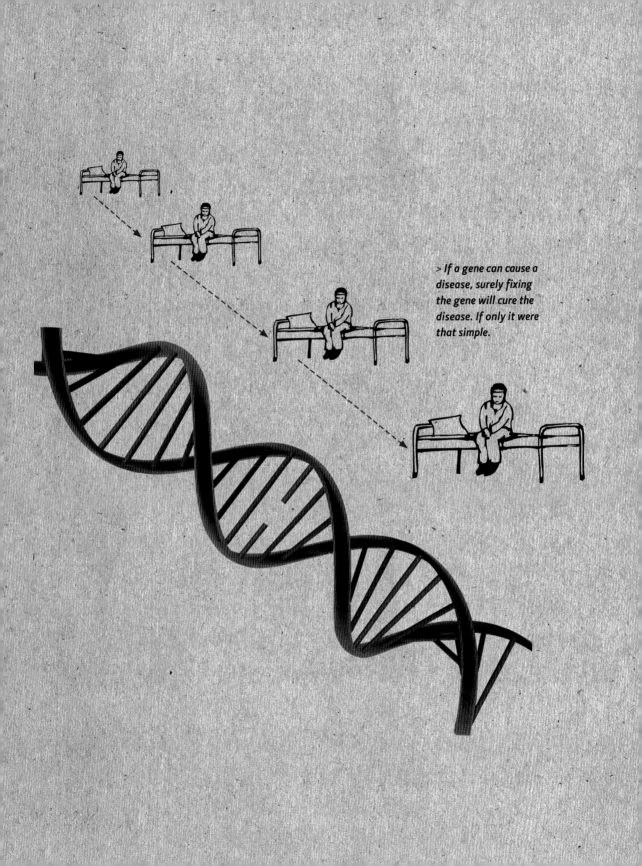

> If a gene can cause a disease, surely fixing the gene will cure the disease. If only it were that simple.

COMPLEMENTARY MEDICINE

the 30-second theory

To its advocates, complementary medicine is exactly what it says: therapy ranging from acupuncture to yoga that can be used alongside conventional medicine. To its detractors, it is a grab-bag of unproven hocus-pocus that occupies patients while conventional medicine does the heavy lifting. They have a point. Many of the conditions that appear to respond well to complementary therapies, such as pain and depression, are those linked to a strong placebo response, in which patients simply think themselves better after receiving otherwise useless therapy. Scientific trials capable of testing such therapies have also given mixed results – not least because many trials are poorly designed. However, there is no denying that there is some evidence that certain forms of complementary medicine, notably acupuncture and meditation, are effective for conditions such as headache, neck pain and stress. And there's certainly no doubting the popularity of complementary medicine. Such therapies have long been widely used in Asian countries, with around 75 per cent of Japanese people regularly using them. Now they are becoming increasingly popular in the West. For example, about one in ten people in Great Britain have used a complementary therapy in the last year.

RELATED THEORIES
see also
THE PLACEBO EFFECT
page 90

3-SECOND BIOGRAPHY
EDZARD ERNST
1948–

30-SECOND TEXT
Robert Matthews

3-SECOND THRASH
The benefits of complementary medicine may be largely in the mind, but who cares, as long as you get better?

3-MINUTE THOUGHT
Sceptics of complementary medicine like to point out that many such therapies have no scientific explanation. However, that is also true of some commonly used conventional medical procedures. For example, there is still no accepted explanation for anaesthesia, whereby certain compounds induce reversible unconsciousness. And you can bet those hard-nosed sceptics would not refuse anaesthesia before surgery, just because they cannot explain why it works.

Complementary medicine is increasingly popular, though in most cases its effectiveness has yet to be proven in large-scale, randomized, controlled clinical tests.

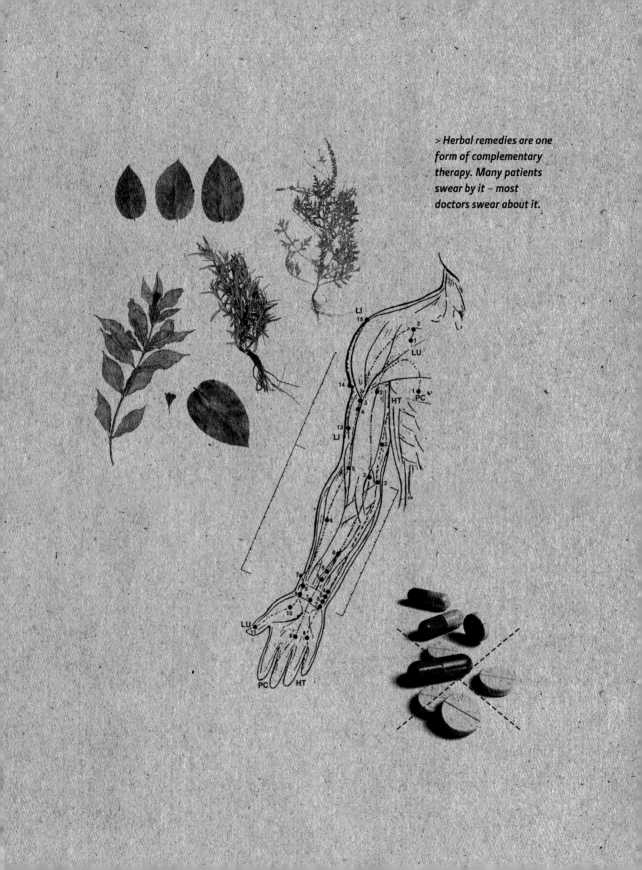

> Herbal remedies are one form of complementary therapy. Many patients swear by it – most doctors swear about it.

EVIDENCE-BASED MEDICINE

the 30-second theory

Once upon a time, doctors decided on treatment for their patients on the basis of personal beliefs – anything from what they heard in medical school years ago to the recommendations of their tennis partners. Sometimes this approach worked – sometimes not. Evidence-based medicine (EBM) seeks to put such decisions on a more scientific footing, based on the results of carefully conducted clinical trials. Doctors can now consult online databases giving the latest thinking about the most effective treatments, based on reviews of clinical trials by internationally respected experts. Whether they actually do or not, however, is another matter. Many doctors still prefer to do their own thing. Their reasons include being too busy to assess the reviews, suspicion about the reliability of clinical trials, and reluctance to being reduced to mindlessly doling out 'approved' therapies. Yet enthusiasts of EBM insist that doctors should always combine the hard evidence with their own judgements about their patients' requirements. In truth, both sides of the argument have a point. Carrying out reviews properly is not easy, and there have been cases of clinical trials being very misleading. Even so, most patients would probably prefer to know their treatment is based on the latest scientific evidence, rather than outdated opinion.

RELATED THEORIES
see also
COMPLEMENTARY MEDICINE
page 86

THE PLACEBO EFFECT
page 90

3-SECOND BIOGRAPHIES
EDZARD ERNST
1948–

ARCHIE COCHRANE
1908–1988

30-SECOND TEXT
Robert Matthews

Doctors all have their favourite therapies and treatments, but what evidence is there that you are getting the best treatment?

> *These ones are always winners!*

THE PLACEBO EFFECT

the 30-second theory

Imagine you are given a pill to take and told it will cure your headache and that your condition then improves, even though the pill contains nothing but chalk. This is the placebo effect. The word means 'I please', but in medicine it refers to treatments that have no real medical effect and work through the power of suggestion. First described in the 1920s, placebos are now an indispensable part of 'evidence-based medicine', the research that finds out whether new drugs or treatments really work. In clinical trials the treatment is compared with a placebo. This might be a pill that looks exactly like the real thing but has no medicine in it, or acupuncture needles that look and feel convincing but do not actually pierce the skin. Typically, in a 'randomized controlled trial', half the patients get the real thing, while the other half receive a placebo. If both groups show the same improvement then you know the new treatment is useless. Placebo effects are extraordinarily powerful and appear to be enhanced by giving bigger pills, pink pills rather than white ones, or by the perceived seniority of the doctor who prescribes it.

RELATED THEORIES
see also
EVIDENCE-BASED MEDICINE
page 88

3-SECOND BIOGRAPHY
ELVIN MORTON JELLINEK
1890–1963

30-SECOND TEXT
Sue Blackmore

A pill does not need to contain any medicine. Thanks to the placebo effect you can believe yourself better – although now you know that, the effect probably won't work so well any more.

> Which one of these pills
will cure your problem?
Who knows? Just pick
one of them and hope
for the best.

PLANET EARTH

cosmic rays This is a stream of radiation and high-energy particles that are spewed out of stars and other bodies, such as quasars. Cosmic rays bombard the Earth from all directions but are filtered out by our magnetic field. The Earth's magnetism directs most of the trapped particles towards the poles, where they interact with gases in the air to form aurorae, such as the Northern Lights.

Earth's crust The rocky outer covering of Earth that, with the uppermost mantle, floats on a partially molten mantle layer (the asthenosphere). The thickest parts of the crust form mountain ranges, the thinnest form the ocean floor.

fixed When a gas from the atmosphere has been absorbed and incorporated into a more complex substance. Most fixation is performed by living things. For example, plants fix carbon dioxide to make sugars during photosynthesis, while some bacteria fix nitrogen to create fertile soils.

feedback mechanism A system that responds to its own activity. A positive feedback mechanism has a runaway effect – activity that reinforces itself so it builds and builds. Negative feedback mechanisms are self-regulating – their activity acts to reduce future activity so that fluctuations are always brought back to a standard level.

glacier A flow of ice that moves slowly across land, generally down from high ground. During ice ages, glaciers covered much of the Earth. Today glaciers are confined to the polar regions and the tops of the highest mountains.

greenhouse gas A gas present in the atmosphere that contributes to the Greenhouse Effect. The most familiar are carbon dioxide and methane, but others include water vapour and CFCs. A greenhouse gas will allow energy from the Sun to reach the Earth's surface and warm it. However, the heat that is then radiated back is blocked from leaving the atmosphere by the gases.

geological Relating to the processes that govern the formation of the Earth's surface. Geologists study mountain building, earthquakes, volcanoes and how different rocks are formed and altered over huge periods of time.

geophysical Relating to an area of geology that uses physics to understand geological processes happening deep within the Earth. These processes cannot be observed directly, so geophysics uses our understanding of magnetism, heat, waves and materials science to create a picture of what is happening.

hypothesis A set of scientific ideas that are proposed to explain a phenomenon observed in nature. A hypothesis is unproven. Should experiments test its ideas sufficiently then a hypothesis is upgraded to a theory – and regarded as true until another theory succeeds in proving it false. The job of scientists is to use theories to formulate new hypotheses and seek out a better version of the truth.

interglacial A period in the Earth's history that falls between two ice ages. All of recorded history has occurred during the current interglacial period.

mantle The hot porridge-like layer of partially molten minerals that extends from the Earth's crust all the way down to the core. The solid rocks of the crust and uppermost mantle float on a plastic layer known as the asthenosphere.

mass extinction An event in which a large percentage of living species becomes extinct in a short space of time. Often entire groups of related species are completely wiped out. The most familiar mass extinction occurred 65 million years ago, when the dinosaurs and many other giant reptiles disappeared forever. No one is quite sure what causes mass extinctions, but it is likely to be the work of extreme natural disasters, such as volcanic eruptions and asteroid strikes.

palaeontologist An expert in fossils, including not just bones and other remains preserved in rocks but all evidence of past life, such as footprints, nesting sites and tools.

parameters The numerical properties, such as temperature and pressure, of the physical phenomenon that a theory describes. The theory is often encapsulated as a mathematical equation linking the parameters.

parts per million A method of expressing very small amounts of a substance mixed into another. As 10 per cent expresses '10 parts per hundred', 5 parts per million shows that there are 5 atoms or molecules of one substance for every million of the other. Parts per billion (ppb) is also used.

salinity A measure of the quantity of salts dissolved in water or another solvent.

uniformitarianism A concept that outlines how the processes that shape the surface of the Earth slowly and relentlessly work at the same rate and in the same way all over the planet, and throughout huge expanses of geological time. As a result we can figure out what happened in the distant past by studying the layers of rocks that cover the Earth today.

SOLAR NEBULAR THEORY

the 30-second theory

The planets formed from a cloud

of particles as diffuse as smoke. This cloud orbited the Sun in a disc, rather like the rings of Saturn. The particles began to stick together because they were all moving the same way around the Sun, but frequently bumped into each other. Eventually the particles grew big enough so that they started to attract one another by gravity, and became bigger still. The largest lumps vacuumed up small lumps to become proto-planets. These larger lumps collided with one another, breaking apart and joining together again many times, before settling down to become the planets as we know them today. Nobody is certain why there are four small, rocky planets near the Sun and four large, gaseous planets further out. It is likely that it was difficult for planets near the Sun to hold on to gases, which were blasted away by the Sun's heat. Conversely, the gas giants were able to collect icy material from beyond the 'frost line' – the distance from the Sun where it is always too cold for the ices to melt – not just water ice, but other frozen substances, such as methane and ammonia.

RELATED THEORIES
see also
RARE EARTH HYPOTHESIS
page 110

THE ANTHROPIC PRINCIPLE
page 122

3-SECOND BIOGRAPHIES
EMMANUEL SWEDENBORG
1688–1772

IMMANUEL KANT
1724–1804

PIERRE-SIMON LAPLACE
1749–1827

30-SECOND TEXT
John Gribbin

3-SECOND THRASH
The planets of the Solar System were born from a cloud of gas and dust left over from the formation of the Sun itself.

3-MINUTE THOUGHT
The planets Uranus and Neptune are so far from the Sun, where the disc of matter from which the planets were formed was thin, that it would have taken hundreds of millions of years for them to form where they are now. It is thought that they formed much closer to the Sun, very near where Jupiter and Saturn are today, and then migrated outwards to their present positions.

From tiny dust particles, mighty planets grow. Planet Earth, the Sun, and the rest of the Solar System were formed from an immense cloud of dust and gas slowly collapsing under the force of its own gravity.

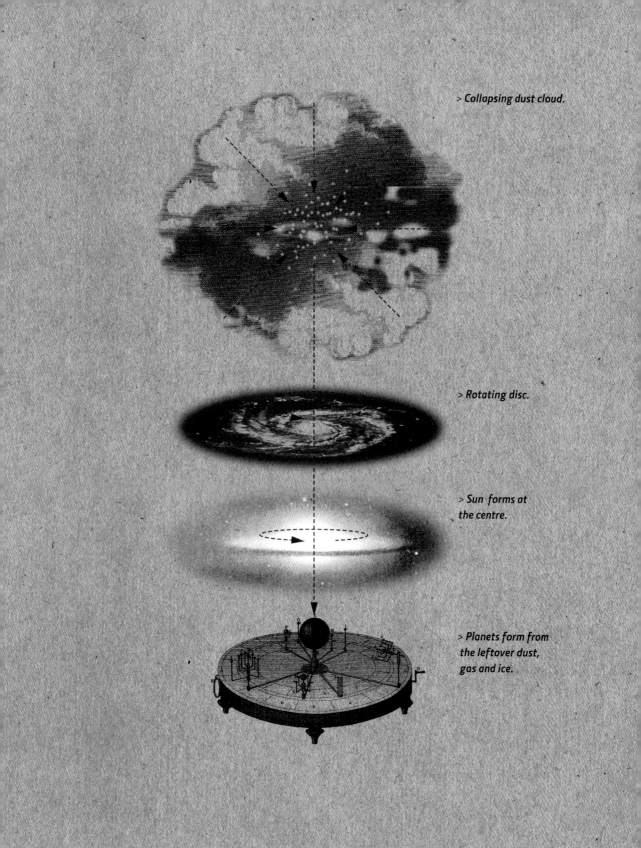

> Collapsing dust cloud.

> Rotating disc.

> Sun forms at
the centre.

> Planets form from
the leftover dust,
gas and ice.

CONTINENTAL DRIFT

the 30-second theory

Our planet is never still, and even the continents themselves move, drifting in a slow dance across the Earth's surface at about the same speed that fingernails grow. Periodically, there are times in our planet's history when this 'dance' brings all the continents clustering together to form a giant supercontinent, before they once again go their separate ways. This happened around 1.3 billion years ago, with the formation of Rodinia, and again 250 million years ago with Pangaea.

The idea of the mobility of continents is actually pretty old hat. As long ago as 1596, noting the congruity of the African and South American coastlines, the Belgian cartographer Abraham Ortelius proposed that the Americas had been violently pushed away from Europe and Africa by earthquakes and floods. Another 300 years were to pass, however, before the German scientist Alfred Wegener came up with the theory that the continents were constantly on the move. Without a mechanism to explain how this could happen, however, the idea was not generally accepted until the early 1960s. By this time new geophysical evidence revealed that continental motion was accomplished by the spreading of the sea floor and driven by currents in a partially molten layer beneath the solid crust and uppermost mantle.

3-SECOND THRASH
While apparently solid and steadfast, the Earth's continents are forever on the move, hitching free rides on the churning convection currents in the porridge-like mantle beneath.

3-MINUTE THOUGHT
Continental drift is now incorporated into plate tectonics. This is the all-embracing model used by geologists to describe how the Earth's crust and uppermost mantle behave. Plate tectonic theory is based on the idea that the Earth's rigid outer layer is made up of more than a dozen large, rocky and constantly moving plates. The theory explains the motions of the continents, where and why earthquakes and volcanoes occur, and how mountain chains are built.

RELATED THEORIES
see also
SNOWBALL EARTH
page 100

3-SECOND BIOGRAPHY
ALFRED WEGENER
1880–1930

30-SECOND TEXT
Bill McGuire

The ground beneath our feet is not completely solid – and it is always on the move.

> *Fancy seeing the world?*
Just stand still, you'll get
all over – eventually.

SNOWBALL EARTH

the 30-second theory

As you might expect, the Snowball Earth theory proposes that way back in deep time glaciers covered the Earth, so that the planet looked like a giant snowball. Proponents of Snowball Earth, including the American geobiologist Joseph Kirschvink, who coined the term, argue that during the appropriately named Cryogenian period, from about 850 to 630 million years ago, a weakly shining Sun, together with a low concentration of greenhouse gases in the atmosphere, led to plunging global temperatures. In response, the Earth became encased in an icy carapace, almost a mile thick.

Unusually low concentrations of greenhouse gases are key to the theory, and ideas abound in relation to how these may have come about. One possibility is that clustering of the continents close to the equator provided superb conditions for rampant tropical weathering, in which carbon dioxide in the atmosphere reacted with rocks to form solid minerals. As a result the concentration of the gas in the atmosphere plummeted. Once the planet was covered in ice, warming up sufficiently to thaw the ice would have proven difficult since the white surface reflected most of the solar radiation back into space. The only ways out would have been an increase in heat from the Sun, or a rise in atmospheric carbon-dioxide levels, perhaps due to the gas venting from volcanoes.

RELATED THEORIES
see also
GLOBAL WARMING
page 104

GAIA
page 108

3-SECOND BIOGRAPHY
JOSEPH KIRSCHVINK
1953–

30-SECOND TEXT
Bill McGuire

3-SECOND THRASH
Brass monkeys would have loved Snowball Earth, an era hundreds of millions of years back in 'deep time', when the planet resembled a frigid ball of ice.

3-MINUTE THOUGHT
Recent research, pointing to the existence of warm periods during Snowball Earth times, shows that climatic cycles were operating that would not have been possible if the Earth had been entirely iced over. It would seem, then, that the extent and degree of the big freeze has been exaggerated, and while glaciation was undoubtedly severe, areas of open, unfrozen ocean must have remained throughout.

The white surface of a frozen Earth would have reflected the Sun's heat back into space – and helped to maintain the freezing temperatures.

> *The Snowball Earth would have sparkled in space as the Sun's light bounced off it.*

1919
Born, Letchworth,
England

1948
Receives a doctorate
in medicine from the
London School of
Hygiene and Tropical
Medicine

1954
Awarded the Rockefeller
Travelling Fellowship
in Medicine; goes to
Harvard University
Medical School

1961
Joins NASA to work on
Surveyor probe to
the Moon

1964
Becomes independent
scientist

1974
Elected fellow of the
Royal Society

1979
Publishes *Gaia*

Planet Earth

JAMES LOVELOCK

James Lovelock is one of the few independent scientists of modern times. He has not worked in a university or government laboratory for 40 years. Often outspoken and always radical, Lovelock is part researcher, part inventor and part visionary. His most famous theory, known as the Gaia Hypothesis, is a description of the way Earth regulates itself. It is not wholly accepted by the scientific establishment, which questions the methods Lovelock used to test his ideas. Although it is a compelling theory, it remains to be seen if Gaia will ever be a useful tool for other scientists.

James Lovelock was born in Letchworth, north of London, in 1919. He studied chemistry in Manchester before joining the National Institute for Medical Research (NIMR) in 1941. Most of his early work at NIMR was related to the war effort. For example, he invented an underwater blood-pressure monitor and a device for measuring the speed of sound. After 20 years at NIMR, Lovelock went to work for NASA. There, his expertise was used to build detectors that could analyze the composition of the rocks and atmosphere on the Moon or a distant planet. Many of the NASA probes equipped with Lovelock's devices were tasked with finding evidence of life. Lovelock realized that the most obvious sign of life on another planet would be its dynamic atmosphere with an ever-changing composition. The atmosphere of a dead planet would just stay the same. Thinking about the way life altered the Earth's atmosphere gave Lovelock the idea for the Gaia theory.

Lovelock left NASA in 1964 and returned to England to become a self-funded scientist and inventor. In 1979, he published his first book, simply titled *Gaia*. This book was soon embraced by the environmentalist community, but was dismissed by many scientists as being driven by a New Age philosophy, rather than objective scientific enquiry. However, many feel that Gaia theory has proved useful, stimulating many years of research and debate. In 2004 Lovelock lived up to his maverick reputation again when he declared that nuclear power was the best way of tackling climate change. This time the criticism has come from the environmentalists.

GLOBAL WARMING

the 30-second theory

Global warming is the somewhat innocuous term for the ongoing and relentless heating up of our world. The Earth is an extraordinarily dynamic planet, with a climate that has experienced wild swings in temperature throughout its 4.6-billion-year history. Currently, we are in a mild interglacial period sandwiched between the last ice age, which ended about 10,000 years ago, and the one to come. Normally during an interglacial, the level of carbon dioxide – the main greenhouse gas that keeps the bitter cold of space at bay by trapping the Sun's heat – is around 280 ppm (parts per million). Now, thanks to the polluting impact of 200 years of industrialization, this has climbed to 385 ppm, and is still rising. While there are sceptics who refuse to assign a human cause to the signs of global warming, the link between carbon-dioxide emissions and planetary heating is not new. As long ago as the 1890s, the Swedish chemist Svante Arrhenius calculated that a doubling of atmospheric carbon dioxide would eventually result in a global temperature rise of around 4°C (7.2°F). So far, our planet has warmed by 0.74°C (1.3°F), and Arrhenius's prediction is on course to become reality by around 2100, bringing a hothouse world of climate chaos and environmental degradation.

RELATED THEORIES
see also
SNOWBALL EARTH
page 100

CATASTROPHISM
page 106

3-SECOND BIOGRAPHY
SVANTE ARRHENIUS
1859–1927

30-SECOND TEXT
Bill McGuire

3-SECOND THRASH
A warmer world might seem like a good thing. But be warned, CO_2 keeps the surface temperature of our sister planet, Venus, at a blistering 483°C (901°F).

3-MINUTE THOUGHT
Global warming is not simply a matter of changes to the climate and ocean circulation. Looking back in time, it appears that previous dramatic rises in our planet's temperature have triggered bouts of geological activity, including volcanic eruptions, earthquakes and underwater landslides. The cause seems to be increased stress and strain within the Earth's crust as a consequence of large, rapid rises in sea level. Our future, then, could be geologically violent, as well as hot.

Some like it hot, but it is looking like global warming will not make the climate warm and sunny, just more extreme. Stand by for high winds and lots and lots of rain.

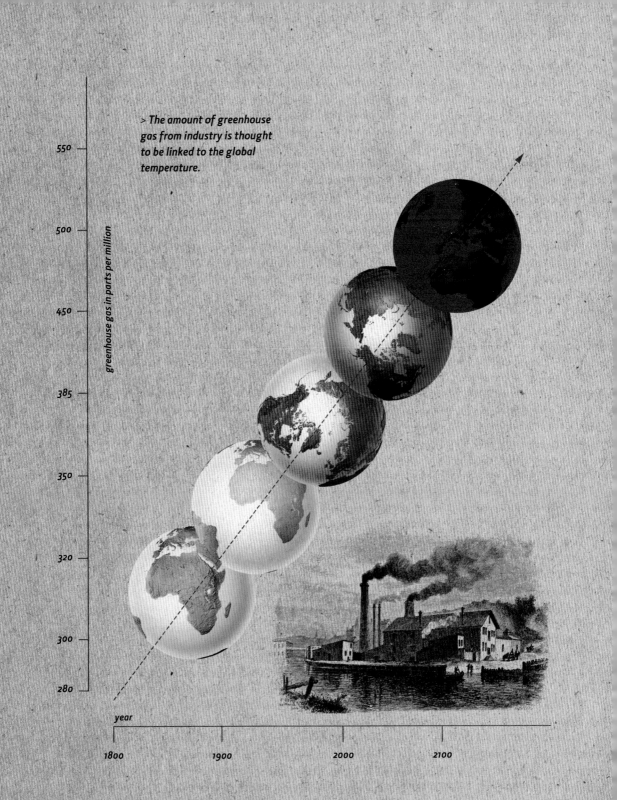

> The amount of greenhouse gas from industry is thought to be linked to the global temperature.

greenhouse gas in parts per million

550

500

450

385

350

320

300

280

year

1800

1900

2000

2100

CATASTROPHISM

the 30-second theory

Catastrophism embraces the idea – much favoured by Hollywood blockbuster movies – that the Earth is periodically afflicted by spontaneous, short-lived and cataclysmic events with global ramifications. This view of the world was driven by the idea that catastrophes were acts of God inflicted upon humans by a vengeful deity, and draws on biblical accounts of doom and disaster, such as Noah's flood. The 18th and 19th centuries marked the heyday of catastrophism as a scientific theory, as natural philosophers argued for a history of the Earth comprising catastrophic events occurring over a period of just a few thousand years. Probably the most famous supporter of catastrophism was the French palaeontologist Georges Cuvier, who linked catastrophes to extinctions that he observed in the fossil record. Since the middle of the 19th century, catastrophism has been largely superseded in scientific circles by uniformitarianism, a concept fathered by the Scottish polymath James Hutton and disseminated by the English geologist Charles Lyell. In stark contrast to catastrophism, this new idea recognizes that Earth history has been largely characterized by gradual, incremental change, involving the same physical processes that we see going on all around us.

RELATED THEORIES
see also
SNOWBALL EARTH
page 100

3-SECOND BIOGRAPHIES
GEORGE CUVIER
1769–1832

JAMES HUTTON
1726–1797

CHARLES LYELL
1797–1875

30-SECOND TEXT
Bill McGuire

3-SECOND THRASH
The history of our planet is punctuated by global cataclysms of epic proportions, compared to which the 2004 Indian Ocean tsunami pales into insignificance.

3-MINUTE THOUGHT
Over the past few decades, catastrophism has made a comeback, as it has become apparent that our planet does succumb periodically to planet-wide cataclysms. So-called global geophysical events (GGEs), or 'gee-gees', such as large asteroid impacts and volcanic super-eruptions, are now known to punctuate our world's uniformitarian calm, triggering mass extinctions, causing worldwide freezes, and bringing the possibility that our civilization may, after all, end with a bang, rather than a whimper.

About 65 million years ago, the dinosaurs and most other giant reptiles died out suddenly. Possible reasons for this catastrophe range from a massive meteor strike, a million-year volcanic eruption and the loss of their fern-tree food.

> *Nothing much happens for 100 million years, then we all get wiped out at once!*

GAIA HYPOTHESIS

the 30-second theory

The Gaia hypothesis was

formulated in the 1960s by the English scientist James Lovelock, who likened the Earth to a self-regulating, living organism. While this does not imply that our world is actually alive, it does entail complex, connected interactions between life and the physical environment – the atmosphere, the oceans, the polar ice sheets and the rock beneath our feet.

According to the Gaia hypothesis, these inter-relationships work in concert to keep the Earth in a moderately stable state that can continue to support life. This balanced condition, sometimes known as homeostasis, is a characteristic of living organisms themselves, which are able to manage their internal processes to maintain the status quo. To bolster his ideas, Lovelock pointed to the extraordinary stability over time of the Earth's surface temperature, despite a progressive rise in solar radiation. He also flagged up the constancy of ocean salinity and atmospheric composition, in the presence of numerous factors that could and should destabilize these parameters. Lovelock's Gaia has been harshly criticized, particularly by biologists such as Richard Dawkins and the late Stephen J. Gould. However, the idea that life plays a crucial role in maintaining the habitability of our world continues to attract both interest and support among serious scientists.

RELATED THEORIES
see also
RARE EARTH HYPOTHESIS
page 110

3-SECOND BIOGRAPHY
JAMES LOVELOCK
1919–

30-SECOND TEXT
Bill McGuire

3-SECOND THRASH
Could it be that our planet is not an inert lump of rock and metal, but something more akin to a self-regulating organism of gigantic proportions?

3-MINUTE THOUGHT
Lovelock, together with his former PhD student Andrew Watson, built a mathematical model known as Daisyworld, to demonstrate how feedback mechanisms could arise in a community of self-interested, living organisms. Daisyworld was populated by black (heat-absorbing) and white (heat-reflecting) daisies in equal number. As the energy output of Daisyworld's Sun was varied, competition between the two types of daisy caused changes in the population balance so as to maintain a temperature close to that needed for optimum daisy growth.

Lovelock's Daisyworld was a simple model of how life on Earth had a regulating effect on the conditions found on the surface of the planet.

> The white daisies like it warm – but they reflect light into space. That makes Daisyworld cool down – perfect conditions for black daisies.

RARE EARTH HYPOTHESIS

the 30-second theory

The Earth may be a rare example

of a home for intelligent life. It has taken more than four billion years of evolution on a quiet, stable planet orbiting a quiet, stable star to produce our civilization. The Sun is an unusually stable star, and it has been a steady source of warmth throughout the evolution of life here.

Earth is largely protected from bombardment by comets by Jupiter; our giant neighbour vacuums up these icy objects before they get to us. Even so, several major terrestrial catastrophes, including the extinction of the dinosaurs, have been linked to cometary impacts. Without Jupiter, these impacts would have been so common that intelligent life would have had no time to evolve on Earth.

Earth's huge Moon is important, too – it stabilizes the axis of the Earth and prevents it from wobbling like a spinning top. The Moon-powered tidal forces inside the Earth keep it hot, and sustain the magnetic field that shields us from harmful cosmic rays. The same forces drive the ocean tides, which played a role in the migration of life onto land. The Moon is believed to be a chunk of the Earth's crust knocked into orbit by an immense impact in the early days of the Solar System. That impact also thinned the Earth's crust, making plate tectonics possible, shifting the continents to and fro, allowing life on Earth to diversify in splendid isolation.

3-SECOND THRASH
We could be the only intelligent life in the entire Universe.

3-MINUTE THOUGHT
The opposite view to the Rare Earth hypothesis is the 'principle of terrestrial mediocrity'. This says that the Earth does not occupy a special place in the Universe – in fact, it is rather normal. There does not seem to be scope in the scientific arguments for any middle ground. So, either life forms like us are common, or we are unique.

RELATED THEORIES
see also
CONTINENTAL DRIFT
page 98

THE ANTHROPIC PRINCIPLE
page 122

30-SECOND TEXT
John Gribbin

There are many happy accidents that have made intelligent life on Earth more likely. Our giant Moon, for example, helps us in numerous ways.

> *The same forces that cause tides also warm Earth's spinning iron core, creating a magnetic field around the globe that deflects cosmic rays.*

> *The Moon-powered ocean tides help drive evolution on Earth.*

THE UNIVERSE

anthropic Relating to humans.

atom The smallest unit of any substance found on the Earth. Atoms themselves are made up of yet smaller particles: protons, neutrons and electrons. The precise combination of these particles gives each type of atom its physical and chemical properties. For example, a gold atom has a different make-up from an atom of carbon. Beyond the Earth and across the Universe, most of the visible matter is made from atoms, but the 'missing' dark matter might be constructed in another way completely.

baryons A family of subatomic particles that includes protons and neutrons. As subatomic particles go, the baryons are the big ones. Smaller particles, such as electrons, photons and quarks, are classed as leptons.

black hole An object in space that forms when the remains of a giant star are compacted into a single point. The gravity of a black hole is so great that nothing, not even light, can escape its pull. Black holes form when the largest stars die.

cosmology The scientific description for the origin and evolution of the Universe. The leading cosmological theory – the Big Bang – as of today, has virtually no rivals.

dimension A fundamental measure used to describe an object or event. Humans are aware of four dimensions – length, width, height and time – but some scientific theories involve multiple dimensions that are only perceived through mathematics.

galaxy A collection of stars that orbits around a central point. The word 'galaxy' is derived from the Greek word for 'milk'. The central core of our galaxy is visible in the night sky as a cloudy strip, dubbed for centuries the Milky Way.

mass A measure of the quantity of matter in an object. 'Mass' and 'weight' are often used interchangeably, but weight is really a measure of the pull of gravity on the object. In everyday terms, the 'mass' and 'weight' of an object are effectively the same on Earth, but on the Moon, the same object's mass is unchanged, while its weight is reduced by 85 per cent in the Moon's lower gravity.

matter The stuff of the Universe, which fills space and can be measured in some way.

neutron star The remains of a dead star that is so densely packed that its protons and electrons have fused to form neutrons. Neutron stars are about as wide as a city, but contain more matter than our Sun.

nucleus The tightly packed centre of an atom, containing particles called protons and, normally, neutrons, too. The protons make the nucleus positively charged, which attracts an equal number of electrons to move around it, thus completing the atom. Almost all the mass of an atom is contained within the nucleus.

quantum A unit that cannot be subdivided any further. Energy exists in quanta.

radiation A term sometimes used to describe the dangerous emissions from radioactive substances, but more correctly used to describe the transfer of photons – tiny packets of energy – through space. Light, heat, radio waves, as well as dangerous gamma rays, are all types of radiation, each of which carry varying amounts of energy. However, more unusual forms of radiation – such as Hawking radiation emitted by black holes are carried by particles of matter.

redshift A phenomenon seen in the light coming from distant stars or galaxies. Distant objects are moving away from Earth – and everything else – as the space-time of the Universe continues to expand. Light waves moving through this expanding space are stretched. This has the effect of increasing the light's wavelength, making it appear redder than it was originally. The increase in wavelength is described as a redshift, even in invisible, colourless forms of radiation. The redshift is one piece of evidence for an expanding Universe. If an object is travelling towards the observer, the opposite happens. The light is compressed and is said to be blueshifted – blue light has a shorter wavelength than red.

subatomic Smaller than an atom.

vacuum A region of space that contains nothing at all, not even invisible gases.

white dwarf The glowing remains left after the death of an average star. Our Sun will eventually become a white dwarf, a small core about the size of Earth. White dwarfs cool gradually into dark 'black dwarfs'. This process is estimated to take 10 billion years, and the Universe is not old enough for any black dwarfs to have formed yet.

THE BIG BANG

the 30-second theory

3-SECOND THRASH
Everything we see around us expanded from a superheated grapefruit, 13.7 billion years ago.

3-MINUTE THOUGHT
The Big Bang raises some questions. How did the Universe begin – where did that grapefruit come from? How will it all end? The first question is answered by the theory of inflation, which explains how a tiny subatomic seed was expanded to the size of a grapefruit by quantum effects. The second question is answered by the recent discovery that the expansion of the Universe is actually getting faster. Therefore the Universe will probably expand forever, with galaxies spread further and further apart in the cosmic darkness.

All the stars we see in the sky are part of a system called the Milky Way galaxy. There are hundreds of billions of stars in the Milky Way, and hundreds of billions of galaxies more or less like it scattered across space, most of them in clusters held together by gravity. A cluster of galaxies is like a huge swarm of bees, moving together as a unit. Studies of the way clusters of galaxies move show that they are getting further apart from one another as time passes. The best evidence for this comes from the redshift in light from distant galaxies. The redshift – literally, a reddening of the galaxies' light – shows that every cluster is getting further away from every other cluster, and there is no centre to the expansion. This is explained by the general theory of relativity, as a result of the space between the clusters stretching. It means that long ago all the galaxies, stars and matter in the visible Universe were piled up in one place, a grapefruit-sized volume of hot energy called the Big Bang. By measuring how fast galaxies are moving apart today, astronomers calculate that the Big Bang occurred 13.7 billion years ago.

RELATED THEORIES
see also
THEORY OF RELATIVITY
page 30

INFLATION
page 120

THE FATE OF THE UNIVERSE
page 130

EKPYROTIC THEORY
page 132

3-SECOND BIOGRAPHIES
GEORGES LEMAÎTRE
1894–1966

ALEXANDER FRIEDMAN
1888–1925

EDWIN HUBBLE
1889–1953

GEORGE GAMOW
1904–1968

30-SECOND TEXT
John Gribbin

Imagine a grapefruit that weighs as much as the whole Universe and is heated to a billion billion billion degrees. Welcome to the Big Bang.

> Expanding superheated grapefruit, anyone?

DARK MATTER & DARK ENERGY

the 30-second theory

RELATED THEORIES
see also
THE FATE OF THE UNIVERSE
page 130

3-SECOND BIOGRAPHIES
FRITZ ZWICKY
1898–1974

VERA RUBIN
1928–

SAUL PERLMUTTER
1959–

30-SECOND TEXT
John Gribbin

3-SECOND THRASH
All the bright stars in all the galaxies make up less than 1 per cent of the mass of the Universe.

3-MINUTE THOUGHT
The only way to do away with the need for non-baryonic dark matter and dark energy would be if our understanding of gravity were wrong. That would mean changing the general theory of relativity. This is very difficult, because any new theory would have to explain all the things that the general theory of relativity explains, and then do something more on top. So far, every time someone has come up with such a new theory, new observations have subsequently ruled it out.

Because the amount of light a star gives out depends on its mass, astronomers can 'weigh' galaxies by measuring how bright they are. Since gravity affects the way things move, astronomers can 'weigh' the Universe by studying the way galaxies move and how fast the Universe is expanding. All the bright stars in all the galaxies add up to less than 1 per cent of the mass needed to explain the way galaxies move and the Universe expands. Calculations of the way atoms – the particles that you and everything in your immediate surroundings are made from – were made in the Big Bang show that there could be about four times as much 'dark' atomic matter in clouds of gas and dust between the stars. This is called baryonic matter because it is made from baryons, large particles found in atoms. Baryonic matter is now thought to make up some 4 per cent of the mass of the Universe. The way galaxies move shows that there is still four or five times more non-baryonic dark matter, made from smaller subatomic particles. However, about 74 per cent of the mass needed to account for the way the Universe expands is still completely missing! The latest thinking is that this missing matter is a form of 'dark energy', which fills space and is making the Universe expand faster.

The light from the Universe's stars is not enough to account for its weight. Most of the stuff in the Universe is too dark to see from Earth – and a lot of it appears to be completely invisible!

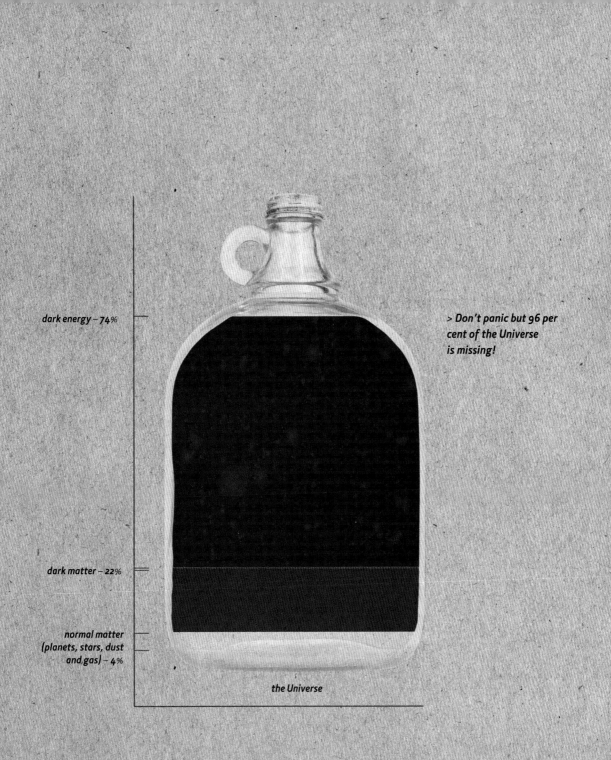

dark energy – 74%

dark matter – 22%

normal matter
(planets, stars, dust
and gas) – 4%

> *Don't panic but 96 per
cent of the Universe
is missing!*

the Universe

INFLATION

the 30-second theory

The Big Bang theory includes an event called inflation. The entire Universe was contained within a single point, before being inflated into an incredibly hot fireball that has since cooled into the galaxies and stars. Quantum uncertainty allows tiny packets of energy to appear out of nothing at all. Usually, these 'vacuum fluctuations' disappear again in a tiny fraction of a second. However, if such a bubble contains a form of energy known as a scalar field, the scalar field can act like antigravity, making the bubble expand extremely rapidly up to a volume about 10 centimetres (4 in) across – roughly the size of a grapefruit – before the field gives up its energy in the form of heat, and the antigravity effect ends.

At the end of this inflation process the Universe is a grapefruit-sized fireball of energy, still expanding, but more sedately, as a result of the push it has received from inflation. This is the Big Bang. Inflation theory predicts that a certain pattern of ripples would be imprinted on space-time at the grapefruit stage. These ripples provided the irregularities from which galaxies and clusters of galaxies could grow by gravitational accretion as the Universe continued to expand. The pattern of galaxies and clusters seen in the Universe today exactly matches the pattern of ripples predicted by inflation.

RELATED THEORIES
see also
QUANTUM MECHANICS
page 38

THE UNCERTAINTY PRINCIPLE
page 40

THE BIG BANG
page 116

EKPYROTIC THEORY
page 132

3-SECOND BIOGRAPHIES
ALAN GUTH
1947–

ANDREI LINDE
1948–

30-SECOND TEXT
John Gribbin

3-SECOND THRASH
In a split second, our Universe sprang from a volume of space one hundred billion billion times smaller than a proton.

3-MINUTE THOUGHT
Inflation explains the Big Bang, but what explains inflation? Can 'nothing at all' really produce vacuum fluctuations? Some cosmologists are looking for something there before inflation, within which the initial fluctuation occurred. This could have been another universe like our own, and if so our Universe could 'give birth' to other universes. The only rival to inflation today is the Ekpyrotic theory in which the Universe is reborn, phoenix-like, out of its own ashes.

Nothing can move faster than light – unless it is space-time itself. During the inflationary period, the Universe's expansion broke the cosmic speed limit.

THE ANTHROPIC PRINCIPLE

the 30-second theory

3-SECOND THRASH
Does the fact that the Universe is just right for life that it was designed with us in mind?

3-MINUTE THOUGHT
The anthropic principle has two forms: weak and strong. In the weak version, the observed values of all physical and cosmological quantities are not equally probable, but take on values that allow the emergence of carbon-based life somewhere in the Universe. The strong anthropic principle states that the Universe *must* allow life to develop within it at some stage in its history.

The Universe we live in is just right for life as we know it. For example, if the strength of gravity were a little greater, stars would be smaller. They would use up their nuclear fuel more quickly and would burn out before there was time for complex life forms, such as ourselves, to evolve. The anthropic principle says that we can use the fact of our existence to predict the value of certain properties of the Universe, such as the strength of gravity. Famously, the astronomer Fred Hoyle used this argument in the 1950s to predict certain properties of the nuclei of carbon atoms, because our form of life depends on carbon, and without those properties carbon could not be made inside stars, and we would not exist. Hoyle's prediction was later confirmed by experiments. The question, then, is why the Universe, like the baby bear's porridge in the *Goldilocks* story, is 'just right'. Some people think it means the Universe was designed for us. Others think it means there must be a multitude of universes – making a Multiverse – and life only exists in the ones like ours.

RELATED THEORIES
see also
RARE EARTH HYPOTHESIS
page 110

3-SECOND BIOGRAPHY
FRED HOYLE
1915–2001

30-SECOND TEXT
John Gribbin

Earth's place in the Universe is like baby bear's porridge – not too hot and not too cold. It is just right. Is this a mere coincidence?

> Like the porridge for
Goldilocks, our Universe
is just right for us.

1942
Born, Oxford, England

1963
Begins research in cosmology; is diagnosed with amyotrophic lateral sclerosis

1974
Proposes concept of Hawking radiation

1979
Appointed Professor of Mathematics at Cambridge University

1988
Publishes *A Brief History of Time: From the Big Bang to Black Holes*

2002
Publishes *The Theory of Everything*

2007
Becomes the first quadriplegic to float in zero gravity

STEPHEN HAWKING

Stephen Hawking has inherited the status of scientific icon from Albert Einstein. Paralysed almost completely by a muscle-wasting disease, Hawking is confined to a wheelchair and communicates using a computer-generated voice. The image of a razor-sharp mind locked inside a helpless body has made Hawking world famous. However, his stature in the scientific world is based on his work in theoretical physics, which has seen him take the academic post at Cambridge University once occupied by Sir Isaac Newton.

Hawking was born in Oxford in 1942. His early academic career at his hometown university was unremarkable. Still in perfect health, the youthful Hawking preferred to socialize than study. In 1962, Hawking moved to do a PhD at Cambridge University. He became ill soon after. It was also at this time that Hawking became interested in cosmology – the study of the origin and evolution of the Universe. Cosmology at Cambridge was to become his life's work.

In the 1970s Hawking became an expert on black holes. Today, most people will have heard of a black hole, and many will know that they are ultra-dense objects that have such strong gravity that nothing, not even light, can escape their pull. However, by 1974 Hawking had already taken this definition to the next level. Black holes contain billions of tons of matter but can be as small as a proton. These massive but also minute objects can be described in terms of both relativity (a theory about big things) as well as quantum mechanics (a theory about small things). Hawking used these two theories to show that black holes actually do release matter in the form of tiny particles – material since named Hawking radiation.

In 1988, Hawking published *A Brief History of Time*, one of the best-selling cosmology books of all time. Since then his celebrity status has seen his iconic voice used in pop music and adverts. More recently, Hawking experienced weightlessness courtesy of a NASA training aircraft, and is planning a space flight.

COSMIC TOPOLOGY

the 30-second theory

What is the shape of the

Universe? Topology is the study of shapes and how one shape can be turned into another without tearing it. It is said that a topologist is someone who cannot distinguish between a doughnut and a coffee cup! If a doughnut was made of rubber you could stretch it into the shape of a cup – the inner ring becomes the cup's handle, and the rest of the doughnut can be shaped into the body of the cup. Most astronomers think that the Universe is infinite. However, if it were finite, it could be shaped like a very big doughnut. If that were the case, by looking one way round the ring of the doughnut you would see the same galaxies you could see by looking the other way round the ring.

In a more complicated topology, imagine a cube in which opposite faces are connected. If you travel in a spaceship up through the 'roof' you would come back into the cube through the 'floor'. Even if the Universe is finite, observations show that it does not have so simple a topology as this. Instead, studies of radiation left over from the Big Bang hint that the Universe may be shaped like a five-dimensional dodecahedron, very similar to the three-dimensional arrangement of pieces that make up a football.

RELATED THEORIES
see also
THE BIG BANG
page 116

3-SECOND BIOGRAPHY
JEAN-PIERRE LUMINET
1951–

30-SECOND TEXT
John Gribbin

3-SECOND THRASH
The Universe may be shaped like a five-dimensional football.

3-MINUTE THOUGHT
If the Universe is finite, some points in space may be repeated on different parts of the night sky, like a hall of mirrors. The effect of this is to make 'ghost images', which appear as matching patterns on different areas of the night sky. The patterns will actually appear as mirror images because some are being seen from the 'front', and others from the 'back'. These patterns have not yet been found, but future space observatories may have the power to detect them.

For topologists, coffee and doughnuts go together especially well. To them, the shape of a doughnut and a coffee cup are the same – only the lengths and angles of their sides vary.

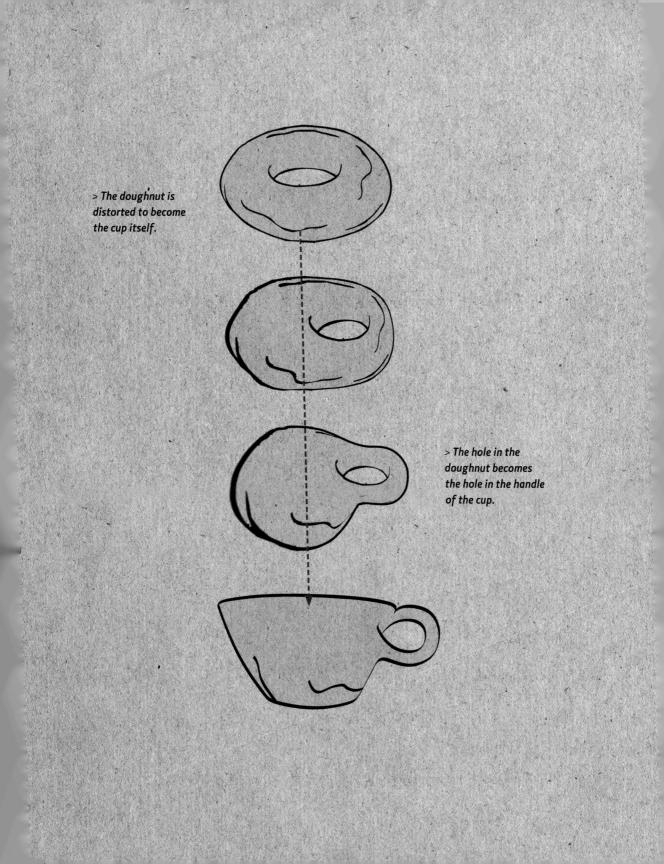

> The doughnut is distorted to become the cup itself.

> The hole in the doughnut becomes the hole in the handle of the cup.

PARALLEL WORLDS

the 30-second theory

Our Universe is thought to have begun when inflation took place in a tiny region of empty space. Even today, dark energy is making the Universe expand faster and faster, and is spreading out the matter in it ever more thinly. Eventually, all that will be left is empty space. Inflation might then produce many more universes in such an empty space. The implication is that our universe is one of many – a Multiverse – produced in a previous phase of a process known as eternal inflation, to which there is no end and no beginning. There may be an infinite number of 'bubble universes' produced in this way. If so, there is no reason to expect the laws of physics to be the same in every universe. Some will be suitable homes for life; others will not. This explains the puzzles of anthropic cosmology. It also means that there will be universes in which every conceivable eventuality actually happens – universes in which you are writing this book and I am reading it, universes where the South won the American Civil War, and universes in which the dinosaurs never died out.

3-SECOND BIOGRAPHIES
HUGH EVERETT
1930–1982

DAVID DEUTSCH
1953–

MAX TEGMARK
1967–

30-SECOND TEXT
John Gribbin

3-SECOND THRASH
Our Universe may be just one of many in an infinite 'Multiverse', in which every possible event is played out somewhere.

3-MINUTE THOUGHT
The Multiverse idea is very similar to the idea of Schrödinger's cat being dead and alive at the same time. One way of explaining this is that there are two universes – in one the cat is dead, while in the other it is alive. This is sometimes called the 'many worlds theory'. The cat exists in both states at the same time, but in different universes.

Are we alone in the Universe? Perhaps, but there are sure to be other earths similar to our own in other universes.

> How many planets like
Earth are out there?

THE FATE OF THE UNIVERSE

the 30-second theory

Dark energy is making the expansion of the Universe accelerate. If this continues – and there is no reason to think it will not – the expansion will get faster and faster as time passes. At first, this will not directly affect matter. Within galaxies, stars will still be born, live out their lives, and die. However, as the stuff from which stars are made is used up, galaxies will get fainter, and more matter will be locked up inside various forms of dead stars – white dwarfs, neutron stars and black holes. But, while this is going on, clusters of galaxies will be moving apart faster and faster, disappearing from each other's view. Our own Milky Way galaxy is part of a small cluster known as the Local Group. Within a couple of hundred billion years – about ten times the present age of the Universe – there will be nothing visible outside this cosmic archipelago. Eventually, space will be expanding so fast that it will overcome gravity and other forces, and material objects will get torn apart. The result will be a rapidly expanding space in which matter has been spread so thin that it is essentially empty – ideal conditions for the birth of one or more new universes through inflation.

RELATED THEORIES
see also
DARK MATTER & DARK ENERGY
page 118

INFLATION
page 120

3-SECOND BIOGRAPHIES
ALEXANDER FRIEDMAN
1888–1925

SAUL PERLMUTTER
1959–

30-SECOND TEXT
John Gribbin

3-SECOND THRASH
The fate of the Universe is to expand faster and faster, until matter is torn apart.

3-MINUTE THOUGHT
Some theorists think that the acceleration of the Universe is itself accelerating. If so, the whole scenario is so speeded up that our Milky Way galaxy will be torn apart in only about 20 billion years from now. This will be just 60 million years before space itself is ripped apart, too. If these theorists are correct, the Universe is already, like a 25-year-old human being, about a third of the way through its life.

It is going to take a few more years yet – 1 with 11 zeros after it, to be precise – but the forces that created the Universe will tear it apart.

> Dark energy will
eventually rip the
Universe apart.

EKPYROTIC THEORY

the 30-second theory

The ekpyrotic theory derives its name from the Greek for 'born in fire'. A better name might be the 'Phoenix Universe'. According to this idea, our Universe is one of a pair of three-dimensional universes, separated from one another by a tiny distance (less than the diameter of an atom) in the fifth dimension. This is actually the fourth dimension of space, but time has already claimed the name 'fourth dimension'. Every point in our space is next-door to a point in the other universe. At present, the two universes are slowly moving apart from one another. In addition, each universe is itself expanding, and so thinning out its contents. Eventually, they will be nothing but empty expanding space. By the time that happens, a spring-like force will be pulling the two universes back together along the fifth dimension. When the two empty universes collide, energy is released and turned into matter, generating a new Big Bang. Because of quantum effects, different parts of the two universes touch each other at slightly different times, making ripples that are the seeds from which galaxies grow. Then, the universes bounce apart and the whole process repeats. It goes on forever. This is the leading alternative explanation for the Big Bang, and it does not require that first phase of inflation.

RELATED THEORIES
see also
THE BIG BANG
page 116

PARALLEL WORLDS
page 128

THE FATE OF THE UNIVERSE
page 130

3-SECOND BIOGRAPHIES
NEIL TUROK
1958–

30-SECOND TEXT
John Gribbin

3-SECOND THRASH
Our Universe may have been born from the collision of two universes moving in the fifth dimension.

3-MINUTE THOUGHT
Surprisingly, the ekpyrotic theory can be tested. Inflation predicts that the Universe should be filled with ripples in space called gravitational waves. The ekpyrotic theory does not predict this phenomenon. Gravitational wave detectors sensitive enough to test the theory will be flown in space in a few years. If they find the waves, it will prove the ekpyrotic theory is wrong. If they don't find the waves, it will prove that inflation is wrong.

Give me your five-dimensional space-time and I'll give you the ekpyrotic Universe in which we yo-yo to and fro with an invisible partner Universe, bouncing in and out of a perpetual sequence of big bangs!

Big Bang

collision point

> There's no peace in an
ekpyrotic Universe – just
a series of continuous
big bangs.

THE KNOWLEDGE

THE KNOWLEDGE
GLOSSARY

algorithm A procedure that is made up of several steps used to find a solution or perform a specific function. Computer programs are algorithmic, in that they are set up to perform a sequence of operations. The word 'algorithm' comes from Al-Khwarizmi, the name of an Arab mathematician from the ninth century AD.

applied mathematics A branch of maths that seeks to build useful mathematical tools that can be applied to solving problems. The alternative field is theoretical maths, which is concerned with the relationships between numbers themselves.

computer model When a computer is used to simulate the real world. Computational science uses these models to study natural processes that cannot be viewed directly, such as one atom bonding to another. Other models are set up to test a system or structure without having to do it for real. The final type of model is used to predict what will happen in the future.

Darwinian evolution Evolution by natural selection, as described by the great naturalist Charles Darwin. The theory states that life forms have changed over the years as they adapt to survive in different conditions. As conditions change, nature selects the life-forms that are best able to survive. The individual life-forms within a population are always slightly different, and these differences are passed from parent to offspring. Some of the population are better at surviving and breeding than others, or, as Darwin would have said, they are fitter. The fitter individuals produce the most young. Each new generation would contain more of the fit individuals, and perhaps eventually all unfit life-forms would be wiped out, unable to compete with their fitter cousins. Evolution by natural selection has taken place.

DNA The abbreviation for deoxyribonucleic acid, a long, chain-like chemical that carries the genetic code.

gene The term gene has two main definitions. The first is a unit of inheritance. This perhaps relates best to the most popular concept of a gene in use today. When people say that they have the gene for red hair, we all understand that they mean they inherited this characteristic from their parents. However, this definition does not tell us much about what physical factors are responsible for giving them red hair. The second definition for gene is a strand of DNA. DNA is the complex chemical that carries the blueprints for a living body in code. DNA is material that is physically replicated to be passed on to the next generation. However, does each gene of DNA relate directly to an inherited characteristic? The answer is rarely, and the relationship between the two concepts is much more complicated than that.

heliocentric Meaning Sun-centred, from the Greek *helios* (Sun), as opposed to geocentric – meaning Earth-centred.

integrated circuit An electronic device in which an entire circuit, with its switches and other components, is made from a single piece of material. The material most often used is silicon; today, tiny integrated circuits are etched on to the surface of a wafer of silicon. Making circuits ever smaller allows an electronic device, such as a computer, to perform more functions at once, and therefore work faster.

law A simple description of a pattern that has been observed in nature. Most laws are expressed as equations.

linear Relating to a straight line. Linear relationships are those that have a direct and unchanging link between the entities involved. If the relationship was represented on a graph, it would form a straight line. Non-linear relationships are very different and harder to represent.

logician Someone who studies the different forms of logic. Logic is a thought system that can be applied to solve problems.

meme An idea, pattern of behaviour, style or belief system that is spread from one person to another within a culture. Meme theorists claim that memes can evolve and mutate in a similar way to biological genes.

replicator An entity that can be passed on from parent to offspring, or that survives the death of its carrier in some other way.

software The instructions given to a computer so it can perform a certain function or solve a particular type of problem. The physical parts of the computer are called the hardware.

teme A meme that has been created and is replicated by technology.

INFORMATION THEORY

the 30-second theory

3-SECOND THRASH
We all rely on information. Information theory shows what it really is, and how to access it as fast and reliably as possible.

3-MINUTE THOUGHT
Information theory may have started out as nerdy maths for engineers, but it has turned out to have deep implications for life, the Universe and almost everything. Biologists have found that DNA incorporates key concepts from the theory to ensure that genes operate correctly, while theoretical physicists have found connections between information theory, black holes and the fundamental laws of physics.

We all think we know what information is, but this theory reveals its true nature – and then shows how to package it up and transmit it as fast and flawlessly as possible. Information theory now underpins a host of everyday technologies, from high-definition digital TV and DVDs to mobile phones and the barcodes on supermarket products. Its origins lie in the work of Claude Shannon, a brilliant young American engineer who developed a precise mathematical theory of information during the 1940s. The most fundamental unit of information he identified was the state of being either true or false – the so-called 'bit', which can take the values of either one or zero. Shannon then used this mathematical definition to reveal a host of insights into how information can be transmitted rapidly yet flawlessly, even in the presence of interference. The results form the basis of so-called 'compression algorithms', which allow entire films to be crammed onto DVDs, or transmitted across the Internet. The theory has also led to so-called error correction codes that keep transatlantic phone-calls crystal-clear – and allow supermarket barcodes to remain readable, even on scrunched-up bags of peanuts.

RELATED THEORIES
see also
QUANTUM ENTANGLEMENT
page 48

3-SECOND BIOGRAPHY
CLAUDE SHANNON
1916–2001

30-SECOND TEXT
Robert Matthews

Information theory can turn a human voice, a picture, or a number into codes of ones and zeros. These codes are easier to store, transmit and copy.

90864678332

>From barcodes to DVDs
– Shannon's mathematical
definition very neatly
quantifies information.

MOORE'S LAW

the 30-second theory

Today's computers make those of just a few years back look like creaky museum pieces. The amount of number-crunching power, memory and hard-disk capacity on our computers all continue to rocket, while prices remain static. This breathtaking rate of improvement was first identified way back in 1965 by Gordon Moore, a co-founder of the microchip maker Intel. In an article in *Electronics* magazine he predicted that the number of electronic components that engineers could cram into each integrated circuit – a rough guide to computing power – would rise within ten years from around 50 to 65,000, roughly equivalent to a doubling every year. By 1975, Moore had modified his prediction to the somewhat more modest rate that now bears his name, which points to a doubling in computing power every two years. Ever since, 'Moore's Law' has proved surprisingly reliable, and is expected to remain so for at least another decade – not least because it has become something that the microchip industry aspires to achieve. Ultimately, however, the laws of physics will ensure this law eventually fails. There is a limit to how small electronic components can be. Moore himself believes the end will come some time around 2025.

RELATED THEORIES
see also
QUANTUM MECHANICS
page 38

3-SECOND BIOGRAPHY
GORDON MOORE
1929–

30-SECOND TEXT
Robert Matthews

3-SECOND THRASH
Always delay upgrading your computer for as long as possible, as the value for money doubles every 24 months.

3-MINUTE THOUGHT
While Moore's Law correctly predicted the increasing computing power of PCs, it failed to predict how this power would be squandered on increasingly bloated software packages. While software designers were once forced to write 'tight code' to suit the small memories of early computers, they now have far fewer constraints – and often leave end-users just as frustrated over the performance of their computer as they were decades ago.

Thinking of buying a computer? Why not wait? They are getting faster, better and cheaper all the time.

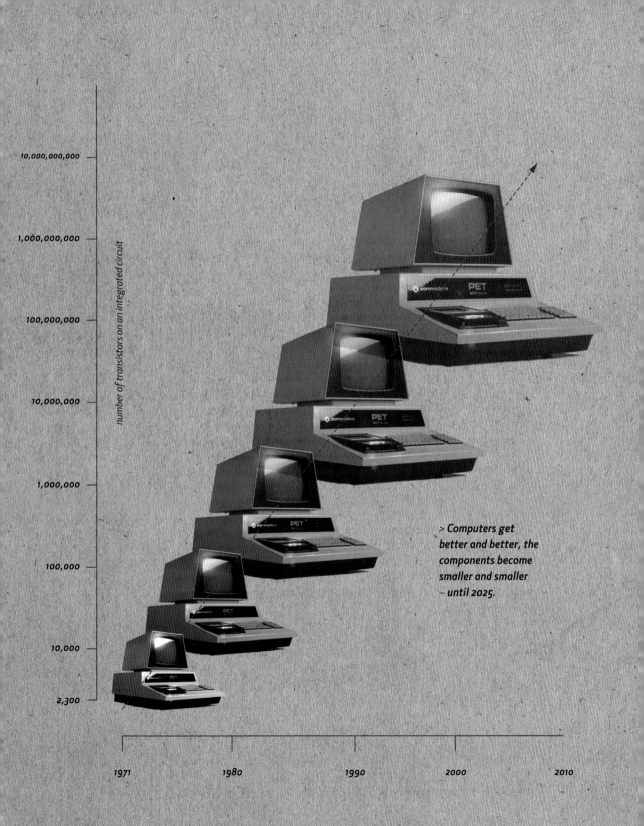

number of transistors on an integrated circuit

10,000,000,000

1,000,000,000

100,000,000

10,000,000

1,000,000

100,000

10,000

2,300

> Computers get
better and better, the
components become
smaller and smaller
– until 2025.

1971   1980   1990   2000   2010

OCKHAM'S RAZOR

the 30-second theory

3-SECOND THRASH
If you have to choose between a neat explanation, making few assumptions, or a messy one relying on stacks of them, go for elegant simplicity every time.

3-MINUTE THOUGHT
As a rule of thumb, Ockham's Razor often works pretty well, but it does not guarantee to identify the correct explanation every time. An example is the conspiracy theory, which often posits incredibly convoluted explanations of historic events. Ockham's Razor suggests we should usually just accept the neat and tidy, official explanation – but only the most gullible would always do so.

There's something about a nice, neat explanation that commands respect. And there is a reason for that. According to a 14th-century English logician named William of Ockham: elegant explanations are more likely to be right than convoluted and messy ones. He recommended making the least number of assumptions needed to do the job when devising explanations – or, as later authors put it, taking a metaphorical razor to them, paring them down to the bare minimum. The underlying motivation is that nature prefers simplicity to complexity. It is not hard to think of examples where this seems to be true. During the Middle Ages, astronomers had been forced to add huge complexity to their Earth-centred view of the Solar System in order to explain the motion of the planets. Yet, simply putting the Sun at the centre made all this complexity redundant. Ockham's Razor suggests this 'heliocentric' view is more likely to be correct – and it did indeed prove to be true. All too often, however, spotting the 'simpler' explanation is easier said than done: for example, is Einstein's law of gravity really simpler than Newton's? Even today, controversy surrounds attempts to turn Ockham's Razor into a rigorous mathematical rule.

RELATED THEORIES
see also
PRINCIPLE OF LEAST ACTION
page 16

3-SECOND BIOGRAPHY
WILLIAM OF OCKHAM
1288–1348

30-SECOND TEXT
Robert Matthews

Keep it simple – that's what's Ockham's Razor is all about. Once you've figured it out, reduce the theory to nothing but its essential elements.

$$\frac{mV_A^2}{2} - \frac{GmM}{(1-\epsilon)a} = \frac{mV_B^2}{2} - \frac{GmM}{(1+\epsilon)a}$$

$$\frac{V_A^2}{2} - \frac{V_B^2}{2} = \frac{GM}{(1-\epsilon)a} - \frac{GM}{(1+\epsilon)a}$$

$$\frac{V_A^2 - V_B^2}{2} = \frac{GM}{a} \cdot \left(\frac{1}{(1'-\epsilon)} - \frac{1}{(1+\epsilon)} \right)$$

$$\frac{\left(V_B \cdot \frac{1+\epsilon}{1-\epsilon}\right)^2 - V_B^2}{2} = \frac{GM}{a} \cdot \left(\frac{1+\epsilon-1+\epsilon}{(1-\epsilon)(1+\epsilon)} \right)$$

$$V_B^2 \cdot \left(\frac{1+\epsilon}{1-\epsilon} \right)^2 - V_B^2 = \frac{2GM}{a} \cdot \left(\frac{2\epsilon}{(1-\epsilon)(1+\epsilon)} \right)$$

$$V_B^2 \cdot \left(\frac{(1+\epsilon)^2 - (1-\epsilon)^2}{(1-\epsilon)^2} \right) = \frac{4GM\epsilon}{a \cdot (1-\epsilon)(1+\epsilon)}$$

$$V_B^2 \cdot \left(\frac{1+2\epsilon+\epsilon^2 - 1 + 2\epsilon - \epsilon^2}{(1-\epsilon)^2} \right) = \frac{4GM\epsilon}{a \cdot (1-\epsilon)(1+\epsilon)}$$

$$V_B^2 \cdot 4\epsilon = \frac{4GM\epsilon \cdot (1-\epsilon)^2}{a \cdot (1-\epsilon)(1+\epsilon)}$$

$$V_B = \sqrt{\frac{GM \cdot (1-\epsilon)}{a \cdot (1+\epsilon)}}.$$

$$\frac{dA}{dt} = \frac{\frac{1}{2} \cdot (1+\epsilon)a \cdot V_B \, dt}{dt} = \frac{1}{2} \cdot (1+\epsilon)a \cdot V_B$$

$$= \frac{1}{2} \cdot (1+\epsilon)a \cdot \sqrt{\frac{GM \cdot (1-\epsilon)}{a \cdot (1+\epsilon)}} = \frac{1}{2} \cdot \sqrt{GMa \cdot (1-\epsilon)(1+\epsilon)}$$

$$T \cdot \frac{dA}{dt} = \pi a \sqrt{(1-\epsilon^2)a}$$

$$T \cdot \frac{1}{2} \cdot \sqrt{GMa \cdot (1-\epsilon)(1+\epsilon)} = \pi \sqrt{(1-\epsilon^2)a^2}$$

$$T = \frac{2\pi \sqrt{(1-\epsilon^2)a^2}}{\sqrt{GMa \cdot (1-\epsilon)(1+\epsilon)}} = \frac{2\pi a^2}{\sqrt{GMa}} = \frac{2\pi}{\sqrt{GM}}\sqrt{a^3}$$

$$T^2 = \frac{4\pi^2}{GM}a^3.$$

$$T^2 = \frac{4\pi^2}{G(M+m)}a^3.$$

MEMETICS

the 30-second theory

Whenever we copy habits, skills, stories, songs or any kind of information from person to person, we are dealing in memes. The idea of memes, along with all scientific theories is itself a meme. The 'meme' meme emerged from the theory of Universal Darwinism; the idea that when any information is copied, varied and selected, then evolution must happen.

Our most familiar replicator is the gene, but in 1976 Richard Dawkins argued that culture consists of a second replicator, and he called this the 'meme'. Humans copy memes (including ideas, skills and behaviours) through imitation and teaching; they vary what they copy by making errors, by deliberate alterations or by creative combining; and they select which memes to remember and pass on. The science of memetics studies how memes spread, why some thrive and others fail, and the consequences of this for the evolution of culture. Generally, some memes spread because they are useful or advantageous to us, like parts of science and medicine, financial institutions, arts and music. Others spread like viruses, even though they may be useless or even harmful to us, including Internet viruses, chain letters, religions and cults, and useless alternative therapies. We humans are meme machines, and the memes use us for their survival.

RELATED THEORIES
see also
NATURAL SELECTION
page 58

THE SELFISH GENE
page 60

3-SECOND BIOGRAPHY
RICHARD DAWKINS
1941–

30-SECOND TEXT
Sue Blackmore

3-SECOND THRASH
Culture evolves because people select memes, just as biology evolves by the selection of genes. We are the meme machines.

3-MINUTE THOUGHT
Memes have been called an 'empty analogy' and a 'meaningless metaphor'. Most biologists deny that memetics is needed to explain the origin of the big human brain, or our peculiar delight in art and music, and argue that existing theories are better. Perhaps memetics is too scary for some – we humans are meme machines, and now that techno-memes (or temes) are developing ever better technology, our role is becoming ever less relevant.

All the ideas in your head are competing with each other. They want you to tell someone else about them. That way they can get into a new head and spread further still.

> You say 'memes', I say
'memetics' – let's pass the
whole thing on.

1928
Born, Bluefield, West Virginia

1945–48
Studies at the Carnegie Institute of Technology in Pittsburgh

1948
Begins doctorate studies at Princeton University

1950
Publishes thesis on game theory

1951
Joins faculty of the Massachusetts Institute of Technology (MIT)

1959
Diagnosed with schizophrenia

1994
Receives Nobel Prize for Economics

2001
A Beautiful Mind, a movie based on Nash's life story, is released

JOHN NASH

It is difficult to understand the
workings of the mind of a mathematician.
After all, he or she is thinking in a language of
numbers, not words. As a result, the work of
math geniuses often becomes sidelined by
the more accessible theories of physical
scientists. However, John Nash is a rare breed
– a famous mathematician.

In the 1950s, Nash explored the ideas
behind zero-sum games – a particular type of
competition in which a gain by one side results
in an equal loss by the other. His ideas were
behind the MAD strategy – mutual assured
destruction – that underwrote the arms race
between East and West during the Cold War.
Nuclear war was averted by the certainty that
the aggressor would come off just as badly
as the defender. Nash's game theory is also
used by economists to predict the behaviour

of markets, and as a result Nash was awarded
the Nobel Prize for Economics in 1994.

John Forbes Nash, Jr., was born in Bluefield,
West Virginia, in 1928. He graduated from the
Carnegie Insitute of Technology in Pittsburgh
in 1948. Two years later he published his thesis
on non-cooperative games. However, his
future career would be a chequered one.
He worked for the RAND Corporation think-
tank and MIT throughout the 1950s, but was
forced to leave following bouts of mental
illness and brushes with the law. In 1959 Nash
began treatment for paranoid schizophrenia.
When well enough, he continued his work on
an informal basis at Princeton University. His
tumultuous life was depicted in the 2001 film
A Beautiful Mind, which won four Oscars.
Perhaps inevitably, the mathematical content
of the film was greatly simplified.

GAME THEORY

the 30-second theory

It's an age-old problem that routinely confronts everyone, from military planners to card-players: What is the best strategy to adopt, given that we do not know what the other guy is thinking? Solving such problems is the *raison d'être* of a branch of applied mathematics called game theory, which, despite its name, has applications far beyond simple pastimes. The first major insight came in the 1920s, when mathematicians devised a rule for dealing with so-called 'zero-sum' games, where one person's gain is exactly matched by the other person's loss. Known as the minimax theorem, it recommends adopting a strategy which gives the biggest payoff in the worst circumstances. However, most real-life 'games' are not zero sum, and some strategies can lead to both sides benefiting or losing. In 1950, the American mathematician John Nash managed to expand the minimax theorem to include these non-zero-sum games as well, greatly extending the usefulness of game theory. For example, evolutionary biologists have used it to understand why animals opt to cooperate rather than fight each other, while psychologists have applied it to the behaviour of criminals in a law-abiding society.

RELATED THEORIES
see also
SOCIOBIOLOGY
page 68

3-SECOND BIOGRAPHY
JOHN NASH
1928–

30-SECOND TEXT
Robert Matthews

3-SECOND THRASH
If you think life is just a game, you need to know how to play it. Game theory can help.

3-MINUTE THOUGHT
For all its apparent sophistication, game theory makes certain assumptions – of which perhaps the most questionable is the rationality of the 'players' taking part. This assumption worked just fine during the Cuban Missile Crisis, as the United States knew that, in the final analysis, the Soviet Union really did not want to be obliterated in a nuclear attack. But when it comes to suicide bombers and the mentally ill, all bets are off.

Everything – from war to business – is a game. And game theory is a type of maths that can help you be a winner.

> It's your move.
Have you thought
this through?

SMALL WORLD HYPOTHESIS

the 30-second theory

3-SECOND THRASH
To make the most of networking, get to know the handful of people who really get around. But think twice before sleeping with them.

3-MINUTE THOUGHT
The small-world effect has its dark side – as vividly demonstrated in August 2003. A power cable touched a tree just outside Cleveland, Ohio, and triggered power cuts that left 50 million people without electricity across eight US states and much of eastern Canada. The loss of power revealed other 'small worlds' linked to the electrical grid, including Canada's air and traffic networks, which were plunged into chaos.

You are talking to a total stranger at a party, discover you have a mutual friend in common and exclaim 'Well, it's a small world!' Indeed it is, and understanding why has given rise to a new scientific discipline known as small world theory, which is providing insights into issues from the spread of disease to the effects of globalization. At its core is the idea of a network made up of interconnected units – anything from friends and neighbours to computers or multinational companies. Such networks tend to be a mixture of short-range connections – for example, families living in a small village – plus a few random, long-range connections, such as those villagers whose jobs have them travelling far and wide. Mathematicians have shown that just a handful of these random links are enough to short-circuit otherwise vast networks, turning them into 'small worlds', where everyone is connected to everyone else by relatively few intermediaries. Indeed, studies suggest that everyone can be linked to everyone else in the world by around six intermediaries. This discovery has focused attention on identifying the people who form the key shortcuts around the networks, because they can hold the key to, for example, the spread of infectious diseases, or the success of a new marketing campaign.

RELATED THEORIES
see also
CHAOS THEORY
page 152

3-SECOND BIOGRAPHY
STANLEY MILGRAM
1933–1984

30-SECOND TEXT
Robert Matthews

In our interconnected world, one tiny event can have far-reaching consequences that affect millions.

> It's a small world – who'd have thought the actions of a hungry mouse could lead to widespread chaos?

CHAOS THEORY

the 30-second theory

You leave the house five minutes late – and miss your train to the airport. Then when you reach the airport, you have missed your flight and discover that the next flight is not until tomorrow. A five-minute overrun has ballooned into a whole day's delay. This is an everyday example of what mathematicians call a non-linear phenomenon, where small effects do not necessarily have small consequences. Chaos theory focuses on such situations, the outcomes of which are often neither totally random, nor totally predictable. One all-too-familiar example is the weather, where non-linear effects ensure small observational errors grow over time to the point where they wreck all hope of making reliable predictions. Forecasters even talk of the 'butterfly effect', where just the flap of a butterfly's wing is enough to make significant changes to any forecast.

Chaos theory provides the tools needed to tell the difference between an apparently random – and thus truly unpredictable – phenomenon, and those that are merely chaotic, with some hope of being correctly forecast. It also gives estimates of the ultimate time scales beyond which no prediction can be trusted – for the weather, it is around 20 days.

RELATED THEORIES
see also
GAIA
page 108

3-SECOND BIOGRAPHIES
HENRI POINCARÉ
1854–1912

EDWARD LORENZ
1917–2008

BENOÎT MANDELBROT
1924–

30-SECOND TEXT
Robert Matthews

3-SECOND THRASH
Death and taxes may be the only certainties in life, but there is a lot of stuff that is not totally random either – it's called chaos.

3-MINUTE THOUGHT
The weather is often taken to be the classic example of chaos in nature, and forecasters have been quick to seize on chaos theory to explain away their less-than-wonderful record of success. There is evidence to suggest that the principal problem may not be chaos, however, but basic flaws in the computer models used to make forecasts.

According to chaos theory, one tiny error can escalate into a major event – even a disaster.

> *In a chaotic world a mis-hit cricket ball could conceivably bring about Armageddon.*

NOTES ON CONTRIBUTORS

EDITOR

Paul Parsons is the former editor of BBC *Focus* magazine. He has written on popular science for publications ranging from the *Daily Telegraph* to *FHM*. His book *The Science of Doctor Who* was one of 12 books longlisted for the 2007 Royal Society Prize for Science Books.

FOREWORD

Martin Rees is President of the Royal Society, Master of Trinity College, and Professor of Cosmology and Astrophysics at the University of Cambridge. He was appointed Astronomer Royal in 1995, and was nominated to the House of Lords in 2005 as a cross-bench peer. He was appointed a member of the Order of Merit in 2007. He has worked and travelled extensively overseas. He has been a Visiting Professor at many universities including Harvard, Caltech, Berkeley, Kyoto and the Institute of Advanced Studies at Princeton where he is now a trustee. He was Regents Fellow of the Smithsonian Institute, Washington, between 1984 and 1988, and is a foreign associate of the National Academy of Sciences, the American Academy of Arts and Sciences, and the American Philosophical Society. He has lectured, broadcast and written widely on science and policy, and is the author of seven books for a general readership.

WRITERS

Jim Al-Khalili is a Professor of Physics at the University of Surrey, where he also holds a chair in the Public Engagement of Science and is an Engineering and Physical Sciences Research Council Senior Media Fellow. Jim is the author of several successful popular science books including *Black Holes, Wormholes and Time Machines* and *Quantum: A Guide for the Perplexed*. In 2007, he received the Royal Society's Michael Faraday medal and prize for science communication. Jim is a regular contributor to radio and television science programmes.

Susan Blackmore is a freelance writer, lecturer, broadcaster and a Visiting Lecturer at the University of the West of England, Bristol. Her research interests include memes, evolutionary theory, consciousness and meditation. She writes for several magazines and newspapers, a blog for the *Guardian*, and is a frequent contributor and presenter on radio and television. Her books include *The Meme Machine* and *Conversations on Consciousness*.

Michael Brooks is a former features editor at *New Scientist* and has written for publications as varied as the *Guardian*, the *Times Higher Educational Supplement*, and *Playboy*. He is the author of two books: *Entanglement*, a novel, and *13 Things That Don't Make Sense*, an exploration of scientific anomalies. He holds a PhD in quantum physics and is a consultant for the magazine *New Scientist*.

John Gribbin is a British science writer and a Visiting Fellow in Astronomy at the University of Sussex. As a science writer, he has written for the science journal *Nature*, the magazine *New Scientist*, *The Times*, the *Guardian* and the *Independent*, as well as their Sunday counterparts and BBC radio. He is best known for his book *In Search of Schrödinger's Cat*, the definitive bluffer's guide to quantum physics. He published his 100th book, *The Fellowship*, in 2005.

Christian Jarrett is staff journalist at *The Psychologist* magazine and editor of The British Psychological Society's *Research Digest*. He has also written for numerous other magazines and organizations including *New Scientist*, *Psychologies*, the Centre for Affective Sciences in Geneva and Unilever. Christian completed his PhD in Behavioural Neuroscience at the University of Manchester. He also has a Masters in Neuroscience from the Institute of Psychiatry in London and a first-class honours degree in Psychology from Royal Holloway, University of London. He is the author of *This Book Has Issues: Adventures in Popular Psychology*.

Robert Matthews is Visiting Reader in Science at Aston University, Birmingham. He has published research on areas ranging from pure mathematics and medical statistics to urban myths such as the origins of Murphy's Law. He is also an award-winning science journalist, his work appearing in publications ranging from *New Scientist*

and *The Financial Times* to *Reader's Digest* and www.thefirstpost.co.uk. He is currently Science Consultant to BBC *Focus* magazine, and the author of *25 Big Ideas: The Science That's Changing our World* and *Why Don't Spiders Stick to their Webs*.

Bill McGuire is Professor of Geophysical Hazards at University College London, and widely accepted as one of the UK's leading hazard specialists. He is also a science writer whose books include *A Guide to the End of the World: Everything You Never Wanted to Know*, *Global Catastrophes: A Very Short Introduction* and, most recently, *Seven Years to Save the Planet*. Bill presented the BBC Radio 4 series *Disasters in Waiting* and *Scientists Under Pressure*, and the Channel 5/Sky News series of short films *The End of the World Reports*.

Mark Ridley was a biology student at Oxford University, and then had spells as a Research Fellow in Oxford and Cambridge Universities. His research specialized in the theory of evolution and animal behaviour. He was a Professor in the Departments of Anthropology and Biology at Emory University, Atlanta, for several years. He returned to the Department of Zoology, Oxford University, as a temporary lecturer, and remains there as a freelance writer. He has written several books, including *Evolution and Mendel's Demon*, many articles and reviews in specialist journals, as well as the *Times Literary Supplement* and daily and Sunday newspapers.

RESESOURCES

BOOKS

25 Big Ideas: The Science That's Changing Our World
Robert Matthews
(Oneworld, 2005)

Chaos: Inventing a New Science
James Gleick
(Penguin, 1988)

Dreams of a Final Theory
Steven Weinberg
(Vintage, 1994)

Gaia: A New Look at Life on Earth
James Lovelock
(Oxford University Press, 2000)

Global Catastrophes: A Very Short Introduction
Bill McGuire
(Oxford University Press, 2006)

Grammatical Man
Jeremy Campbell
(Simon & Schuster, 1982)

Prisoner's Dilemma
William Poundstone
(Anchor, 1993)

Quantum: A Guide for the Perplexed
Jim Al-Khalili
(Weidenfeld & Nicolson, 2004)

Seven Years to Save the Planet: The Questions and Answers
Bill McGuire
(Weidenfeld & Nicolson, 2008)

Six Degrees: The Science of a Connected Age
Duncan Watts
(W. W. Norton & Company, 2004)

Snowball Earth
Gabrielle Walker
(Three Rivers Press, 2004)

Supercontinent: 10 Billion Years in the Life of Our Planet
Ted Nield
(Granta Books, 2007)

Understanding Moore's Law: Four Decades of Innovation
Edited by David C. Brock
(Chemical Heritage Foundation, 2006)

What We Believe But Cannot Yet Prove
John Brockman
(Harper Perennial, 2006)

MAGAZINES/ARTICLES

Focus
www.bbcfocusmagazine.com

New Scientist
www.newscientist.com/home.ns

Wired
www.wired.com/

Anderson, M. C. and Green, C. (2001),
'Suppressing unwanted memories by
executive control.'
Nature, 410, 131–134.
www.nature.com/nature/journal/v410/
n6826/full/410366a0.html

Solms, M. (2004), 'Freud returns.'
Scientific American, 290, 82–88.
www.sciam.com/article.cfm?id=freud-
returns-2006-02

WEBSITES

Bad Science
www.badscience.net
*Ben Goldacre's column from the Guardian,
presented as a weblog. Articles generally
focus on how the media misrepresents
science*

Genetics Education Center
www.kumc.edu/gec/
Online genetic medicine resource

Information theory
www.tinyurl.com/f4two
Claude Shannon's original paper

*The International
Neuropsychoanalysis Centre*
www.neuropsa.org.uk/npsa/

The James Lind Library
www.jameslindlibrary.org
*Evidence-based medicine
online resource*

Null Hypothesis
www.null-hypothesis.co.uk
*The journal of unlikely science –
a lighthearted look at the weird
world of science and technology*

Open2.net
www.open2.net/alternativemedicine/
index.html
*Online resource for complementary
medicine*

Stanford Encyclopedia of Philosophy
plato.stanford.edu/entries/simplicity/
entry on Ockham's Razor

INDEX

ACKNOWLEDGEMENTS

PICTURE CREDITS

The publisher would like to thank the following individuals and organizations for their kind permission to reproduce the images in this book. Every effort has been made to acknowledge the pictures, however we apologize if there are any unintentional omissions.

Corbis: 8, 43, 124, 146
Getty Images: 7, 22, 51, 62
Science Photo Library: 44, 82, 102.